The Many Faces of RNA

The Many Faces of RNA

Edited by

D. S. EGGLESTON

SmithKline Beecham Pharmaceuticals
New Frontiers Science Park
Harlow, Essex, UK

C. D. PRESCOTT

SmithKline Beecham Pharmaceuticals
709 Swedeland Road
King of Prussia, PA, USA

and

N. D. PEARSON

SmithKline Beecham Pharmaceuticals
New Frontiers Science Park
Harlow, Essex, UK

ACADEMIC PRESS

San Diego   London   Boston   New York
Sydney   Tokyo   Toronto

ACADEMIC PRESS INC
525B Street
Suite 1900
San Diego, CA 92101, USA

ACADEMIC PRESS LIMITED
24–28 Oval Road
LONDON
NW1 7DX, UK

A catalogue record for this book is available from the British Library
ISBN 0-12-233210-5

Typeset by Laser Words, Madras, India
Printed and bound in Great Britain by
MPG Books Limited, Bodmin, Cornwall

98 99 00 01 02 03 EB 9 8 7 6 5 4 3 2 1

Contents

Contents

Contributors

R. Brimacombe Max-Planck-Institut für Molekulare Genetick, AG-Ribosomen, Ihnestrasse 73, 14195-Berlin, Germany

S. Cusack European Molecular Biology Laboratory, Grenoble Outstation, c/o ILL, 156X, F-38042 Grenoble Cedex 9, France

D. E. Draper Department of Chemistry, Johns Hopkins University, Baltimore, MD 21218, USA

M. J. Gait Medical Research Council, Laboratory of Molecular Biology, Hills Road, Cambridge CB2 2QH, UK

S. T. George INNOVIR Laboratories Inc., 510 E. 73rd Street, New York, NY 10021, USA

A. R. Goldberg INNOVIR Laboratories Inc., 510 E. 73rd Street, New York, NY 10021, USA

S. M. Hecht Departments of Chemistry and Biology, University of Virginia, Charlottesville, Virginia 22901, USA

A. Jäschke Institut für Biochemie, Freie Universität Berlin, Thielallee 63, D-14195 Berlin, Germany

L. A. Kirsebom Department of Microbiology, Box 581, Biomedical Center, S-751 23 Uppsala, Sweden

M. Ma INNOVIR Laboratories Inc., 510 E. 73rd Street, New York, NY 10021, USA

B. Metcalf SmithKline Beecham Pharmaceuticals, 709 Swedeland Road, King of Prussia, PA 19406, USA

B. Polisky NeXstar Pharmaceuticals, Inc., 2860 Wilderness Place, Boulder, Colorado 80301, USA

C. D. Prescott SmithKline Beecham Pharmaceuticals, 709 Swedeland Road, King of Prussia, PA 19406, USA

J. D. Puglisi   Center for Molecular Biology of RNA, University of California, Santa Cruz, CA 95064, USA

B. S. Sproat   INNOVIR GmbH, Olenhuser Landstrasse 20B, 37124 Rosdorf, Germany

G. Varani   MRC Laboratory of Molecular Biology, Hills Road, Cambridge CB2 2QH, UK

Introduction

The Many Faces of RNA, Cambridge, March 1997

The Many Faces of RNA served as the subject for the eighth in a series of SmithKline Beecham Pharmaceuticals Research Symposia, held at Robinson College, Cambridge in March 1997. Our primary purpose in sponsoring these symposia, in alternate years between the United States and the United Kingdom, is to highlight a rapidly developing area of scientific investigation with the potential to impact the discovery of therapeutic modalities in a heretofore underappreciated or underdeveloped way. These gatherings are conceived as a vehicle for focus on an area of scientific investigation which, for any number of reasons, may only now be achieving a level of interest and support commensurate with increased attention by the pharmaceutical industry as a whole and SmithKline Beecham particularly. In short, a field which is 'hotting up'. The style and format are deliberately designed to promote in-depth presentations and discussions and to facilitate the forging of collaborations between academic and industrial partners. In addition, we hope that the symposium will benefit the student community by exposing a current theme of interest to the pharmaceutical industry, thereby highlighting it as a field ripe for future study. This book gathers together written versions of the oral presentations, which comprised the bulk of the meeting contents, and abstracts for the plethora of poster presentations which complemented those efforts and further fostered interaction between participants.

Previous topics in the series have included 'Chirality in Drug Design', 'Genomes, Molecular Biology and Drug Discovery' 'Neurodegeneration' and 'Organometallic Reagents in Organic Synthesis', among others. Traditionally the themes have alternated between a biological and a chemical topic; however, in designing this year's symposium it was clear that a blend of the two was the only way forward. We attempted to be broad in the scope of science reviewed, in keeping with the myriad functions of RNA, but were unescapably constrained by the amount of time available and breadth which could have been covered. Thus some current and highly relevant topics had to be excluded out of sheer necessity (for example, antisense technology, the fascinating phenomena of RNA editing and processing, the use of the unnatural enantiomer of RNA to facilitate the design of diagnostic tools and biological screens and the whole area of RNA virology). Instead we chose to combine presentations which would offer a fuller appreciation for the structural complexities inherent in RNA-based systems with several that examined the use of RNA as a tool in exploring biological systems and the development of interesting chemistry in the belief that such a combination offered the potential to generate insights applicable to the process of target selection and the design/discovery of novel agents intended to interact with those targets of therapeutic value. Overall, the opportunity for each speaker was to present specific state-of-the-art results from their recent work within the context of an overview of how that work enhances understanding of RNA structure and function. Within those illustrations an emphasis on the importance of close collaboration between biochemists, chemists and biologists in driving the exploration and development of new methodologies was clearly apparent. The nature of these interactions forms a strong basis for the discovery of keys for understanding the potential for broader application within the chemical and pharmacological context.

This year's topic was one most appropriate to timely examination within the splendid setting at Cambridge University where so much forefront work on RNA and its many functions has been carried out over the years and continues to this day. As a brief example we note that a laboratory in Cambridge was among the earliest (1968), if not the first, to report crystallization of a tRNA. Indeed, in 1974 the first full three-dimensional structure determination of a tRNA was produced at Cambridge (Aaron Klug) and in 1976 the first structure determination of a tRNA synthetase appeared from the MRC laboratory headed by David Blow. From these beginnings to the most recent exciting determinations of hammerhead ribozyome structure and of RNA–protein complexes through the application of modern NMR and crystallographic studies in the areas of HIV and spliceosome activities, Cambridge remains a focal point for development of investigational methods for RNA and RNA-dependent phenomena.

RNA is a molecule of many facets and subtleties, many of which are only now beginning to be fully appreciated and exploited. Like any topic of current interest, an appreciation of RNA, the evolutionary molecule, is evolving both as a tool for increased understanding of biological systems and, perhaps more importantly, as a target for therapeutic intervention. Thus gaining a fuller appreciation of the structural forms which RNA may adopt and the molecular basis and structural complexity of RNA–protein recognition (which bears many resemblances to protein–protein interaction), and of the interactions between RNA and 'small molecule' ligands, will serve as a platform for advancing drug discovery. Diversity of RNA function requires a diversity in presentation of tertiary structural features which in turn facilitates the recognition of three-dimensional surfaces and charge distributions unique to different RNAs – in contrast to DNA recognition where different principles apply.

The principal tools for such investigations are NMR, electron microscopy (for large assemblies) and X-ray crystallography but, in the past, progress along these lines has been slow despite some intense effort. However, these approaches are coming into their own as applied to RNA-based systems due to technological improvements and to advances in the fields of RNA synthesis and purification leading to amounts of material in a form required for structural studies. Traditional techniques such as chemical footprinting continue to play a fundamental role. Likewise approaches to the isolation and study of ribosomal proteins are beginning to blossom. Novel techniques, such as the use of modified bases to incorporate deuterium and thereby remove proton spectral overlap, and four-dimensional NMR spectroscopy are driving forward structural studies in solution. The engineering of sequences to impart structural stability while preserving critical features has also begun to impact success. Convertible nucleosides and related modifications will undoubtedly find wide future utility in rigidifying structures for crystallographic examination and providing sites for heavy atom incorporation to assist phase determination.

In addition, there is still much to be learned from nature herself since the site of action for many known (and probably currently unknown) naturally occurring antibiotics is RNA or RNA–protein complexes. Thus the development of a fuller appreciation of these in a biochemical and structural sense can potentially lead to the engineering of improved activities through the design of new agents. Starting with microbiological targets, and extending to virology and perhaps transcriptional control within eukaryotic host cells, the future of RNA-based research as applied in the pharmaceutical industry context appears very bright and exciting indeed.

Finally, RNA is often regarded as a prime candidate for 'The Original Living Molecule'. As such there is every possibility that it will become recognized as having basic functions in the cell for which mechanisms are currently unknown. Its already recognized widespread roles in processes as fundamental as information transfer, protein synthesis, self-editing and gene regulation are sufficient to keep many laboratories occupied for the forseeable future, but newly implicated roles, for example, as a timing mechanism for gene expression, give further impetus to the overall fascination. Many facets, many faces, much yet to learn about a subject rich in activity, history, information, structure, and potential.

The pharmaceutical industry is constantly seeking to understand new, directed targets focused on crucial phenomena which can be exploited for disease prevention and cure. Recent visionary efforts to utilize the wealth of data from genome sequencing efforts for the discovery of novel protein targets for therapeutic intervention can be complimented, particularly in the first instance in the antibacterial and antiviral areas, through the application of modern molecular and cellular techniques to the increased understanding of RNA and RNA–protein complexes as targets for drug interaction. The extremely powerful SELEX techniques used in both *in vitro* and *in vivo* contexts demonstrate the utility of RNA as a tool for probing biochemistry and potentially for developing diagnostic and therapeutic RNA species. In addition, the catalytic activity of RNA enzymes may find utility in previously unexplored biological and chemical settings such as for catalysis of organic reactions or for diagnostic purposes. It remains to be seen how these efforts can be translated into practical applications in the therapeutic arena and therefore how many other companies follow SmithKline Beecham's lead in pharmaceutical research directed toward RNA-based targets, as has been the case with the genomics efforts initiated several years ago.

We hope the symposium and book will have helped stimulate progress in translating potential advances and approaches in the study of RNA with an endpoint of exploiting RNA in drug discovery and biomedical applications. We extend a special thank-you to the speakers for their oral and written contributions, and to so many of the participants who contributed to the science over those two days through display and discussion of posters.

We wish to express out sincere appreciation to Beatrice Leigh, Sam Luker and Aisling Dolly for their expert efforts in organizing this year's symposium. Three cheers to all! Also, many thanks to Jacqui Kinnison for her huge effort in pulling together all of the annotations and getting all of those full-stops and semi colons correct.

I

OVERVIEW

1

RNA as a Therapeutic Target for Bleomycin

SIDNEY M. HECHT

Departments of Chemistry and Biology, University of Virginia, Charlottesville, Virginia 22901, USA

Abstract

Small molecule–DNA interactions have been studied intensively for many years, and a number of these molecules find utility as antitumour agents. In contrast, until recently much less emphasis had been given to the targeting of RNA. This has undoubtedly been due in part to the greater technical difficulties inherent in the preparation and handling of RNA, and probably also to the more recent realization that RNA constitutes a potential therapeutic target.

There are a number of attractive features of RNA as a therapeutic target, including (i) the substantial number of secondary and tertiary structures accessible to RNA, arguing for potential selectivity of drug–RNA interaction, (ii) the accessibility of many RNAs within the cytoplasm of cells, (iii) the less intensive packaging of RNA, as compared with DNA, and (iv) the paucity of cellular mechanisms for RNA repair. That targeting of RNA can lead to alteration of cell function is illustrated by the actions of ricin, a protein that kills cells by creating a lesion in ribosomal RNA and onconase, an antitumour protein with ribonuclease activity believed to function at the level of tRNA inactivation.

The potential of small molecules to function therapeutically at the level of RNA degradation is illustrated by bleomycin, a clinically important antitumour antibiotic long believed to function at the level of DNA degradation. Recent evidence indicates that bleomycin is also capable of effecting RNA strand scission. All major classes of RNA can be cleaved by Fe•bleomycin, including tRNAs and tRNA precursor transcripts, mRNAs, rRNAs and RNA–DNA heteroduplexes. However, not all RNAs in any of these classes are cleaved by Fe•bleomycin; even those RNAs that do act as substrates are not all cleaved with the same efficiency. Cleavage of substrate RNAs typically also occurs at a smaller number of sites than DNAs of comparable size.

THE MANY FACES OF RNA
ISBN 0-12-233210-5

Bleomycin has also been shown to effect the cleavage of *E. coli* dihydrofolate reductase mRNA within intact *E. coli* cells. That RNA may constitute an additional therapeutic locus for the expression of antitumour activity by bleomycin is supported by the accessibility of RNA within the cell cytoplasm. The less intensive packaging of RNA, as compared with DNA, and the paucity of mechanisms for RNA repair also argue for the relevance of RNA as a therapeutic target for bleomycin.

1 Introduction

The bleomycins (BLMs) (Fig. 1) are antitumour antibiotics originally isolated from *Streptomyces verticillus* (Umezawa *et al.*, 1966a,b). They are used extensively in the clinic, principally for the treatment of squamous cell carcinomas

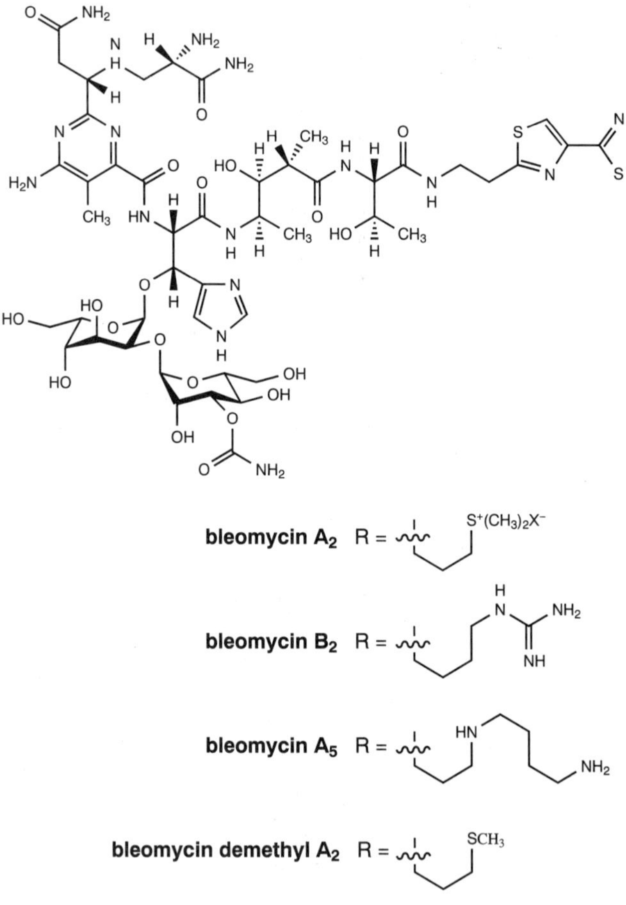

bleomycin A₂ R =

bleomycin B₂ R =

bleomycin A₅ R =

bleomycin demethyl A₂ R =

Fig. 1 Structures of several bleomycin group antitumour antibiotics.

and malignant lymphomas (Carter *et al.*, 1978; Sikic *et al.*, 1985). While early mechanistic studies led to the suggestion of a number of possible targets for bleomycin, including DNA ligase, DNA and RNA polymerases, and DNA and RNA nucleases (Tanaka *et al.*, 1963; Falaschi and Kornberg, 1964; Mueller and Zahn, 1976; Ohno *et al.*, 1976), the focus of most studies in recent years has involved the ability of bleomycin to cleave and otherwise degrade DNA, a process that obtains both *in vitro* (Hecht, 1986; Stubbe and Kozarich, 1987; Natrajan and Hecht, 1993) and *in vivo* (Barranco and Humphrey, 1971; Hittelman and Rao, 1974; Barlogie *et al.*, 1976).

Bleomycin-mediated DNA degradation involves both a metal ion cofactor and dioxygen; a number of redox-active metal ions can support the degradation of DNA by BLM but Fe, and possibly Cu (Ehrenfeld *et al.*, 1987), must function in this capacity in a therapeutic situation. DNA degradation by BLM is oxidative in nature and affords two sets of products, both of which involve the initial formation of C-4′ deoxyribose radicals and have been characterized explicitly at a chemical level. DNA cleavage by BLM is sequence-selective, involving predominantly 5′-GT-3′ and 5′-GC-3′ sequences; both single- and double-strand breaks are produced (Hecht, 1986; Stubbe and Kozarich, 1987; Natrajan and Hecht, 1993). As a consequence of the clinical utility of bleomycin and its highly distinctive mechanism of DNA recognition and degradation, considerable efforts have been expended to identify other agents that function at the level of DNA damage. It has seemed reasonable to anticipate that some of these might also prove to have useful antitumour activities.

In addition to its DNA damaging properties, bleomycin has at least two other biochemical effects that might be thought to contribute to its antitumour activity. The first is its ability to effect lipid peroxidation (Gutteridge and Fu, 1981; Ekimoto *et al.*, 1985; Nagata *et al.*, 1990; Kikuchi and Tetsuka, 1992), presumably using the same reactive intermediates that lead to DNA damage. That lipid peroxidation may actually represent an additional therapeutic locus for BLM was suggested by the finding that a BLM analogue incapable of DNA cleavage nonetheless inhibited the growth of a mammalian cell line when the culture medium also contained the local anaesthetic dibucaine (Berry *et al.*, 1985b).

Another cellular target for BLM that is of potential interest therapeutically is RNA. While early studies failed to detect RNA binding (Hori, 1979) or cleavage by BLM (Suzuki *et al.*, 1970; Haidle *et al.*, 1972; Müller *et al.*, 1972; Haidle and Bearden, 1975; Hori, 1979), subsequent studies have shown unequivocally that RNA cleavage can obtain under conditions comparable to those noted for DNA cleavage by BLM (Magliozzo *et al.*, 1989; Carter *et al.*, 1990).

There are a number of reasons to think that RNA might constitute an additional therapeutic locus for the expression of antitumour activity by bleomycin. These include the accessibility of much cellular RNA within the cell cytoplasm, the less intensive packaging of RNA as compared with DNA, and the apparent paucity of mechanisms for repair of damaged RNA.

It is instructive to consider a few basic facets of RNA cleavage by BLM, which in the aggregate may provide some perspective on the issue of RNA as a therapeutic target for bleomycin. These include the spectrum of RNA structures susceptible to degradation by BLM and the characteristics associated with cleavage of those RNAs. The ability of BLM to effect RNA degradation within an intact cell is obviously critical to the use of this macromolecule as a therapeutic target. Finally, it is instructive to consider the evidence that cellular RNA damage, mediated by any mechanism, can actually lead to cell death.

2 RNAs that are cleaved by bleomycin *in vitro*

The three major classes of cellular RNAs include transfer RNAs (tRNAs), ribosomal RNAs (rRNAs) and messenger RNAs (mRNAs). As illustrated below, while not all RNA molecules are substrates for cleavage by Fe•BLM, there are examples of RNAs in each of these major classes that do act as substrates.

The most extensively studied substrates to date have been tRNAs and tRNA precursor transcripts. Yeast cytoplasmic tRNA^{Asp} is cleaved by Fe(II)•BLM A$_2$ at two sites, G$_{45}$ and U$_{66}$ (Fig. 2) and the tRNA^{Asp} shown at the right side of the figure is also a substrate for cleavage by Fe•BLM. However, in spite of the fact that these two species must have rather similar three-dimensional shapes as they are both substrates for activation by yeast aspartyl-tRNA synthetase (Perret *et al.*, 1990), the sites of Fe•BLM-mediated cleavage differ. The analogous result was obtained for three additional tRNA^{Asp} transcripts; although none was terribly

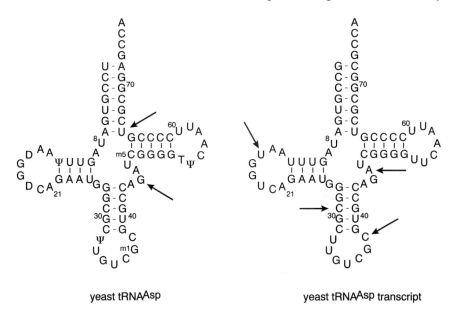

Fig. 2 Sites of cleavage (denoted by arrows) of mature yeast cytoplasmic tRNA^{Asp} (left) and a tRNA^{Asp} precursor transcript (right).

different in primary sequence from the species shown in Fig. 2, the sites of Fe•BLM-mediated cleavage differed quite substantially (Holmes *et al.*, 1996). The logical conclusion to be drawn from these observations is that the sites of tRNA (transcript) cleavage by Fe•BLM must be very sensitive to changes in tRNA structure.

A survey of tRNAs and tRNA transcripts revealed a number that were substrates for cleavage by Fe•BLM and others that were not. The species that served as substrates included *B. subtilis* tRNA^{His} precursor (Carter *et al.*, 1990, 1991a,b), yeast tRNA^{Phe} (Hüttenhofer *et al.*, 1992; Holmes *et al.*, 1993), *E. coli* tRNA$_1^{His}$ and a *Schizosaccharomyces pombe* tRNA^{Ser} suppressor transcript (Holmes *et al.*, 1993). In comparison, *E. coli* tRNA^{Tyr} precursor (Carter *et al.*, 1990) and *E. coli* tRNA^{Cys} (Holmes *et al.*, 1993) were not substrates for cleavage by Fe•BLM under any tested condition; neither were yeast mitochondrial tRNA^{Asp} nor tRNA$_f^{Met}$ precursors (Holmes *et al.*, 1993).

The general pattern among those species that served as substrates was that cleavage occurred at fewer sites than would have been expected for a DNA duplex of comparable length, and involved some 5'-G•pyr-3' cleavage sites. However, cleavage also obtained at sequences not normally cleaved in DNA and failed to occur at many other 5'-G•pyr-3' sites. Interestingly, on the assumption that all of the substrates existed as cloverleaf secondary structures (cf Fig. 2), a number of sites of cleavage were at the junctions between single- and double-stranded regions of the RNA.

It is important to note that not all of the tRNAs that acted as substrates for Fe•BLM were cleaved with comparable efficiencies. Among the tRNAs and tRNA transcripts studied, the most efficient substrate was clearly *B. subtilis* tRNA^{His} (Carter *et al.*, 1990; Holmes and Hecht, 1993; Holmes *et al.*, 1993). The efficiency of cleavage of this substrate at low concentrations of Fe(II)•BLM A$_2$ was comparable to that of duplex DNA.

Another efficient substrate for cleavage by Fe•BLM was yeast 5 S rRNA. This RNA was cleaved at three sites by Fe(II)•BLM A$_2$ (Fig. 3), these were all found to be at the uridine nucleotides of 5'-GUA-3' sequences, all of which also had a one-nucleotide bulge one or two nucleotides to the 3'-side of the cleavage site. As first noted by Hüttenhofer *et al.* (1992), the facility of RNA cleavage by Fe•BLM can be diminished substantially by Mg^{2+}. From the perspective of the relevance of RNA as a therapeutic target for BLM, it is important to note that the cleavage of yeast 5 S rRNA was diminished, but not eliminated, by Mg^{2+} concentrations significantly in excess of those that obtain physiologically. The same was also found to be true for Fe•BLM-mediated cleavage of *B. subtilis* tRNA^{His} (Holmes *et al.*, 1993).

The use of rRNAs ^{32}P-labelled alternately at the 5' and 3'-ends also permitted the analysis of the chemistry of cleavage by Fe•BLM; this proved to be analogous to the cleavage of DNA at the level of resolution permitted by polyacrylamide gel analysis (Holmes *et al.*, 1993). In addition, the fact that the two end-labelled substrates each gave the same three cleavage products (cf left and right panels in

(a)

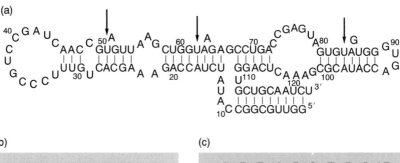

(b)   (c)

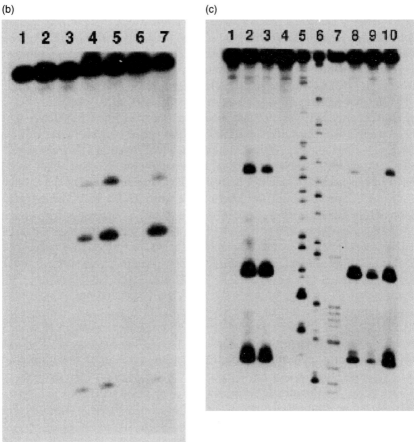

Fig. 3 Fe(II)•BLM A₂-mediated cleavage of yeast 5 S ribosomal RNA; the three sites of cleavage are shown by arrows (top). The left panel illustrates the Fe•BLM-mediated cleavage of 3′-³²P end-labelled 5 S rRNA. Lane 1, rRNA alone; lane 2, 250 μM BLM; lane 3, 250 μM Fe²⁺; lane 4, 250 μM Fe(II)•BLM; lane 5, 210 μM Fe(II)•BLM; lane 6, 210 μM Fe(II)•BLM + 85 mM NaCl + 17 mM Mg²⁺; lane 7, 210 μM Fe(II)•BLM + 85 mM NaCl. The right panel shows the cleavage of 5′-³²P end-labelled 5 S rRNA by Fe•BLM. Lane 1, rRNA alone; lane 2, 250 μM Fe(II)•BLM; lane 3, 125 μM Fe(II)•BLM; lane 4, alkali-treated rRNA; lane 5, G lane; lane 6, A > G lane; lane 7, U + A lane; lane 8, 250 μM Fe(II)•BLM + 100 mM NaCl; lane 9, 250 μM Fe(II)•BLM + 5 mM Mg²⁺; lane 10, 250 μM Fe(II)•BLM + 1 mM Mg²⁺. The rRNA was used at a final nucleotide concentration of ∼1 μM.

Fig. 3) indicates that no double-strand cleavage occurred and that all three sites represent primary sites of cleavage in this RNA (Holmes *et al.*, 1993).

Also cleaved by Fe•BLM was *E. coli* 5 S rRNA. Interestingly, this species exhibited a different pattern of cleavage than was evident following its incorporation into a DNA–RNA heteroduplex (*vide infra*).

As shown for a portion of HIV-1 reverse transcriptase mRNA (Fig. 4), Fe(II)•BLM can also effect the cleavage of messenger RNAs. Although not

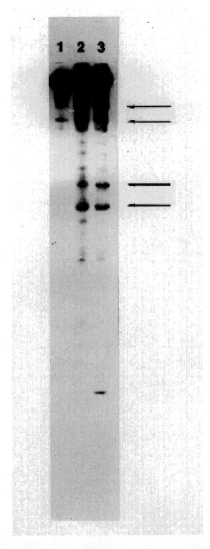

Fig. 4 Fe(II)•BLM-mediated cleavage of an RNA transcript corresponding to the 5′-end of HIV-1 reverse transcriptase mRNA. The 5′-^{32}P end-labelled RNA was treated with no BLM (lane 1), 100 μM Fe(II)•BLM A$_2$ (lane 2) or 500 μM Fe(II)•BLM A$_2$ (lane 3).

cleaved with particularly high efficiency, at least four sites of cleavage were readily apparent in this RNA (Carter *et al.*, 1990; Hecht, 1994). It has also been found that Fe•BLM can cleave bullfrog ferritin mRNA within the iron regulatory element (Dix *et al.*, 1993) and can also effect the degradation of *E. coli* dihydrofolate reductase mRNA (*vide infra*).

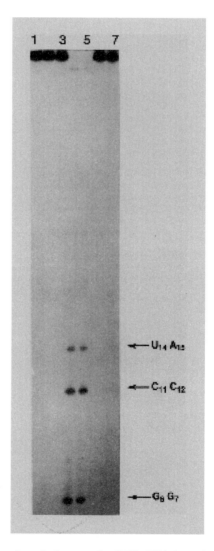

Fig. 5 Three of the four sites of cleavage of a DNA-RNA heteroduplex by Fe(II)•BLM A_2. The 120-base pair heteroduplex was $5'-^{32}P$ end-labelled on the RNA strand and then treated with Fe•BLM. Lane 1, heteroduplex alone; lane 2, 5 µM BLM A_2; lane 3, 5 µM Fe^{2+}; lanes 4-7, 10, 5, 2.5 and 1 µM Fe(II)•BLM. A fourth cleavage band, closest to the $5'$-end of the RNA, migrated off the bottom of this gel during electrophoresis.

3   Cleavage of a DNA–RNA heteroduplex by Fe•Bleomycin

In addition to the three major classes of cellular RNAs, DNA–RNA heteroduplexes are ubiquitous cell constituents, as they are formed as intermediates
during both DNA transcription and reverse transcription of RNA into DNA. A
DNA–RNA heteroduplex formed from *E. coli* 5 S rRNA by reverse transcription was shown (Fig. 5) to be an exceptionally good substrate for cleavage by
Fe(II)•BLM. Cleavage occurred at four sites on the RNA strand, as summarized
in Fig. 6. The cleavage at these sites was apparent even when Mg^{2+} was at mM
concentration (Morgan and Hecht, 1994).

Figure 6 also illustrates the cleavage of the DNA strand of the heteroduplex
by Fe•BLM in direct comparison with cleavage of a DNA duplex 36-base pairs
in length and having the same sequence as the 5′-end of the DNA strand of
the heteroduplex. Interestingly, the three sites cleaved on the DNA strand of the
heteroduplex were identical with three of the five sites cleaved on the analogous
strand of the DNA duplex. Given the substantial sensitivity of BLM cleavage
patterns to even small changes in substrate structure (*vide supra*), this finding
suggests strongly that the DNA strand of the heteroduplex has a structure quite
similar to the analogous strand in the B-form duplex.

4   Structural factors controlling the cleavage of DNA and RNA

While the cleavage of DNA and RNA clearly are different in many ways,
the results obtained with yeast ribosomal 5 S rRNA and with the DNA–RNA
heteroduplex underscore some essential similarities between cleavage of these
two species. Because these two RNA substrates are also arguably the closest in
secondary structure to B-form DNA, these findings raised the issue of whether the
characteristics unique to DNA and RNA cleavage are primarily a consequence
of polynucleotide secondary structure, or of the constituent mononucleotides
that constitute DNA and RNA. Accordingly, we prepared a 'tDNA' molecule
having the same primary sequence as *B. subtilis* tRNA[His] precursor (Fig. 7). As
noted above, this tRNA precursor transcript is a particularly efficient substrate
for Fe(II)•BLM; it is cleaved predominantly at uridine$_{35}$ (Carter *et al.*, 1990).
There was, *a priori*, no way to know whether the tDNA would fold in the same
fashion as tRNA[His] precursor and, even if it did, no reason to expect that it
would necessarily be cleaved in the same fashion. Nonetheless, as summarized
in Fig. 7, the two species actually were both cleaved predominantly at position
35 (Holmes and Hecht, 1993).

One particularly interesting facet of this experiment became apparent when
higher concentrations of Fe(II)•BLM were employed for cleavage of the tRNA[His]
and tDNA[His]. While the tDNA was cleaved at many additional sites and degraded
completely, cleavage of the tRNA did not increase significantly as the concentration of Fe(II)•BLM was increased beyond 2.5 μM. Titration of radiolabelled
tRNA[His] with increasing concentrations of unlabelled tRNA[His] or tDNA[His] was

DNA 5'-ATGCCTGGCAGTTCCCTACTCTCGCATGGGGAGACCCACACTACCATGGCGCTACGGCGTTCACTTCTTGAGTTCGGCATGGGGTCAGGTGGGACCACCGGCGCTACGGCGCCAGGCA-3'
RNA 3'-UACGGACCGUCAAGGGAUGAGCGUACCCUGGGUGUGAUGGUAGCCGCGAUGCCGCAAAGUGAAGACUCAAGCCGUACCCCAGUCCACCCUGGUGCGCGAUGCCGCGCGCGGUCCGU-5'

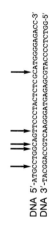

DNA 5'-ATGCCTGGCAGTTCCCTACTCTCGCATGGGGAGACC-3'
DNA 3'-TACGGACCGTCAAGGGATGAGAGCGTACCCCTCTGG-5'

Fig. 6 Sites of Fe(II)•BLM A₂-mediated cleavage of a DNA–RNA heteroduplex and a 36-base pair DNA duplex.

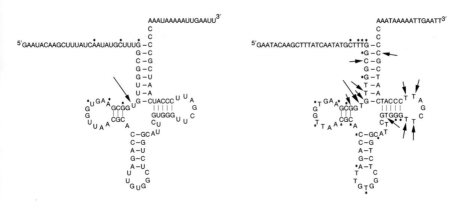

Fig. 7 Structures of *B. subtilis* tRNA^{His} and its corresponding tDNA; both structures are shown folded in the cloverleaf secondary structure believed to be characteristic of tRNAs. Shown in the figure are the major sites of cleavage at U_{35} (T_{35}) (large arrow), other significant sites of cleavage of the tDNA (small arrows) and minor sites of cleavage (asterisks).

compared with the results of a similar titration of radiolabelled tDNA^{His}. Remarkably, the tRNA^{His} clearly bound Fe•BLM more avidly than the tDNA, arguing that the lesser cleavage of the tRNA must be intrinsic (Holmes and Hecht, 1993). This difference could be due to the binding of Fe•BLM to tRNA^{His} precursor in a fashion that does not permit the formation of RNA lesions, or else afford lesions that do not lead to strand scission products. One possible way to accomplish the latter, as recently pointed out by Crich and Mo (1997), would be for the C-4 ribose radicals putatively produced from RNA to have greater stability than those formed from DNA by Fe•BLM.

5 Cleavage of RNA by bleomycin in an intact cell

While the degradation of chromatin in mammalian cells by bleomycin is well established (Barranco and Humphrey, 1971; Hittelman and Rao, 1974; Barlogie *et al.*, 1976), there are no reports of the cleavage of RNA within an intact cell by bleomycin. Presumably this reflects a number of technical difficulties inherent in this type of experiment, including the low concentration of any specific RNA in a cell, the transient nature of many cellular RNAs and the difficulty in assuring that the RNA cleavage product observed after BLM treatment is not simply the transcription product of a damaged DNA.

In spite of these technical difficulties, we have now demonstrated the cleavage of dihydrofolate reductase mRNA in intact *E. coli* by BLM. The design of the experiment is shown in Fig. 8. *E. coli* cells were transformed by a high copy number plasmid containing the gene for dihydrofolate reductase under the control of an inducible promoter. After induction of DHFR mRNA synthesis, the cells were treated with BLM, then lysed and combined with a primer complementary to the 3′-end of the mRNA. Reverse transcription then afforded cDNAs that

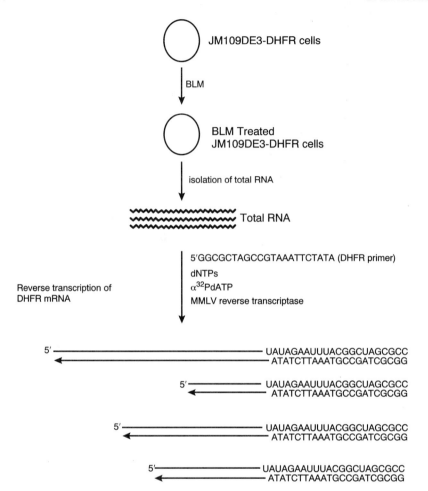

Fig. 8 Scheme illustrating the strategy for analysis of cleavage of DHFR mRNA within intact *E. coli* by bleomycin.

reflected the length of the mRNA following BLM treatment. Because the primer is complementary to the 3'-end of the mRNA, only mRNAs that had been fully transcribed can be observed by the analytical procedure employed. The region of the DHFR mRNA studied was cleaved at three sites in a cell-free system; one of these was cleaved strongly and another moderately strongly in the intact *E. coli* (Snow, 1995).

While the analogous experiment has yet to be carried out in a mammalian cell, this experiment provides unequivocal evidence that BLM can mediate RNA cleavage in a cellular environment. Further, just as the presence of Mg^{2+}

suppressed BLM cleavage at a number of sites that were cleaved with only moderate efficiency, it would seem that within an intracellular environment the selectivity of RNA may be further enhanced. The reason(s) for this additional selectivity have not been investigated directly but could well be due to protein binding by the RNA.

6 The case for RNA as a therapeutic target for BLM

As noted above, there is unequivocal evidence for the cleavage of DNA *in vitro* and *in vivo* by bleomycin and it seems clear that BLM exerts its antitumour effects at least in part at this locus. However, there are some observations which suggest that damage to DNA may not constitute the sole therapeutic locus for BLM; these might be thought to include the poor correlation between DNA damage and growth inhibition induced by BLM in cultured KB cells and the apparent ease with which DNA damage is repaired in such cells (Berry *et al.*, 1985a).

Poddevin *et al.* (1991) have also shown that introduction of BLM into the cytosol of cells by electropermeabilization significantly enhanced the cytotoxicity of BLM, suggesting that the cell membrane, and presumably also the nuclear membrane, may constitute a significant barrier to BLM. In this context RNA represents a particularly attractive target for BLM in that all three major classes of cellular RNAs are present in the cytoplasm where they should be more accessible to BLM than nucleic acids in the nucleus. Since we have shown that all three classes of RNA can be cleaved by BLM *in vitro*, and that at least one mRNA can be cleaved in an intact prokaryotic cell, the main issue would seem to be whether RNA damage can lead to cell death.

It may be pertinent to note that ricin effects the destruction of eukaryotic cells by the depurination of a single nucleotide within the cellular 28 S rRNA (Endo *et al.*, 1987) and that onconase, a ribonuclease with significant antitumour activity, is believed to exert its antitumour effects at the level of tRNA degradation (Lin *et al.*, 1994). Since the production of a number of proteins is tightly linked to mRNA levels, it would not be difficult to imagine that BLM, through mRNA cleavage, could also reduce the levels of some protein required for cell viability.

7 Conclusion

Bleomycin has been shown to cleave all major classes of RNA *in vitro*, as well as DHFR mRNA within intact *E. coli*. In addition to its role as a prototype for antitumour agents that function at the level of DNA cleavage, BLM may also serve as a prototype for new classes of therapeutic agents that function via RNA cleavage.

Acknowledgement

I thank Dr Christine Debouck, SmithKline Beecham Pharmaceuticals, for the plasmid from which HIV-1 reverse transcriptase mRNA was transcribed.

References

Barlogie, B., Drewinko, B., Schumann, J. and Freireich, E. J. (1976). *Cancer Res.* **36**, 1182–1187.
Barranco, S. C. and Humphrey, R. M. (1971). *Cancer Res.* **31**, 1218–1223.
Berry, D. E., Chang, L.-H. and Hecht, S. M. (1985a). *Biochemistry* **24**, 3207–3214.
Berry, D. E., Kilkuskie, R. E. and Hecht, S. M. (1985b). *Biochemistry* **24**, 3214–3219.
Carter, B. J., de Vroom, E., Long, E. C., van der Marel, G. A., van Boom, J. H. and Hecht, S. M. (1990). *Proc. Natl. Acad. Sci. USA* **87**, 9373–9377.
Carter, B. J., Holmes, C. E., Van Atta, R. B., Dange, V. and Hecht, S. M. (1991a). *Nucleosides Nucleotides* **10**, 215–227.
Carter, B. J., Reddy, K. S. and Hecht, S. M. (1991b). *Tetrahedron* **47**, 2463–2474.
Carter, S. K., Crooke, S. T. and Umezawa H., eds (1978). *Bleomycin: Current Status and New Developments*, Academic Press, New York.
Crich, C. and Mo, X.-S. (1997). *J. Am. Chem. Soc.* **119**, 249–250.
Dix, D. J., Lin, P.-N., McKenzie, A. R., Walden, W. E. and Theil, E. C. (1993). *J. Mol. Biol.* **231**, 230–240.
Ehrenfeld, G. M., Shipley, J. B., Heimbrook, D. C., Sugiyama, H., Long, E. C., van Boom, J. H., van der Marel, G. A., Oppenheimer, N. J. and Hecht, S. M. (1987). *Biochemistry* **26**, 931–942.
Ekimoto, H., Takahashi, K., Matsuda, A., Takita, T. and Umezawa, H. (1985). *J. Antibiot.* **38**, 1077–1082.
Endo, Y., Mitsui, K., Motizuki, M. and Tsurugi, K. (1987). *J. Biol. Chem.* **262**, 5908–5912.
Falaschi, A. and Kornberg, A. (1964). *Fed. Proc.* **23**, 940–945
Gutteridge, J. M. C. and Fu, X.-C. (1981). *FEBS Lett.* **123**, 71–74.
Haidle, C. W. and Bearden, J., Jr. (1975). *Biochem. Biophys. Res. Commun.* **65**, 815–821.
Haidle, C. W., Kuo, M. T. and Weiss, K. K. (1972). *Biochem. Pharmacol.* **21**, 3308–3312.
Hecht, S. M. (1986). *Acc. Chem. Res.* **19**, 383–391.
Hecht, S. M. (1994). *Bioconjugate Chem.* **5**, 513–526.
Hittelman, W. N. and Rao, P. N. (1974). *Cancer Res.* **34**, 3433–3439.
Holmes, C. E. and Hecht, S. M. (1993). *J. Biol. Chem.* **268**, 25909–25913.
Holmes, C. E., Carter, B. J. and Hecht, S. M. (1993). *Biochemistry* **32**, 4293–4307.
Holmes, C. E., Abraham, A. T., Hecht, S. M., Florentz, C. and Giegé, R. (1996). *Nucleic Acids Res.* **24**, 3399–3406.
Hori, M. (1979). In *Bleomycin: Chemical, Biochemical and Biological Aspects* (S. M. Hecht, ed.) pp. 195–206, Springer-Verlag, New York.
Hüttenhofer, A., Hudson, S., Noller, H. F. and Mascharak, P. K. (1992). *J. Biol. Chem.* **267**, 24471–24475.
Kikuchi, H. and Tetsuka, T. (1992). *J. Antibiot.* **45**, 548–555.
Lin, J.-J., Newton, D. L., Mikulski, S. M., Kang, H.-F., Youle, R. J. and Rybak, S. (1994). *Biochem. Biophys. Res. Commun.* **204**, 150–162.
Magliozzo, R. S., Peisach, J. and Ciriolo, M. R. (1989). *Mol. Pharmacol.* **35**, 428–432.
Morgan, M. and Hecht, S. M. (1994). *Biochemistry* **33**, 10286–10293.
Mueller, W. E. and Zahn, R. K. (1976). *Prog. Biochem. Pharmacol.* **11**, 28–47.

Müller, W. E. G., Yamazaki, Z., Breter, H.-J. and Zahn, R. K. (1972). *Eur. J. Biochem.* **31**, 518–525.

Nagata, R., Morimoto, S. and Saito, I. (1990). *Tetrahedron Lett.* **31**, 4485–4488.

Natrajan, A. and Hecht, S. M. (1993). In *Molecular Aspects of Anticancer Drug–DNA Interactions* (S. Neidle and M. Waring, eds) pp. 197–242, MacMillan, London.

Ohno, T., Miyaki, M., Taguchi, T. and Ohashi, M. (1976). *Prog. Biochem. Pharmacol.* **11**, 48–58.

Perret, V., Garcia, A., Grosjean, H., Ebel, J.-P., Florentz, C. and Giegé, R. (1990). *Nature* **344**, 787–789.

Poddevin, B., Orlowski, S., Belehradek, J., Jr. and Mir, L. M. (1991). *Biochem. Pharmacol.* **42**, S-66–S-75.

Sikic, B. I., Rozencweig, M. and Carter, S. K., eds (1985). In *Bleomycin Chemotherapy*, Academic Press, New York.

Snow, A. M. (1995). Ph:D. thesis, University of Virginia.

Stubbe, J. and Kozarich, J. W. (1987). *Chem. Rev.* **87**, 1107–1136.

Suzuki, H., Nagai, K., Akutsu, E., Yamaki, H., Tanaka, N. and Umezawa, H. (1970). *J. Antibiot.* **23**, 473–480.

Tanaka, N., Yamaguchi, H. and Umezawa, H. (1963). *J. Antibiot.* **16A**, 86–91.

Umezawa, H., Maeda, K., Takeuchi, T. and Okami, Y. (1966a). *J. Antibiot.* **19A**, 200–209.

Umezawa, H., Suhara, Y., Takita, T. and Maeda, K. J. (1966b). *J. Antibiot.* **19A**, 210–215.

II

RNA: CHEMISTRY, STRUCTURE AND INTERACTIONS

2

Synthetic RNA Modification and Cross-linking Approaches towards the Structure of the Hairpin Ribozyme and the HIV-1 Tat Protein Interaction with TAR RNA

MICHAEL J. GAIT, DAVID J. EARNSHAW, MARK A. FARROW AND NIKOLAI A. NARYSHKIN

Medical Research Council, Laboratory of Molecular Biology, Hills Road, Cambridge CB2 2QH, UK

Abstract

New crystallographic and NMR techniques have provided the first models of biologically interesting RNAs and RNA–protein interactions. However, there is also an important need for complementary RNA techniques not only as tests of these early models but also in studies of the many RNAs that have not yet proved fully tractable to high resolution structural studies. One approach to probing RNA structure and function involves the chemical synthesis of model RNAs carrying individual functional group modifications at key locations thought to be involved in RNA catalysis and in interaction with proteins. In our laboratory these techniques were applied initially to the hammerhead ribozyme and to the HIV-1 Tat protein–TAR RNA interaction. More recently we have applied functional group modification techniques to the two essential loop domains of the hairpin ribozyme. These methods have proved helpful in determining which functional groups are involved in structurally or functionally important hydrogen bonding or other ionic interactions.

THE MANY FACES OF RNA
ISBN 0-12-233210-5

We have now begun to develop new chemical cross-linking approaches aimed at obtaining more three-dimensional information about RNA–RNA and RNA–protein interactions. One project is directed towards understanding the folding of the two helical domains of the hairpin ribozyme. We have determined the ribozyme cleavage parameters of synthetic hairpin ribozymes which have been cross-linked via the 2′-positions of specific nucleoside residues with an aryl or alkyl disulfide linker and compared them with their reduced, unconstrained counterparts. In a second project directed towards obtaining specific cross-links between the model HIV-1 TAR RNA duplex carrying site-specific modifications and a synthetic Tat peptide, we have developed a chemical method for the introduction of a trisubstituted pyrophosphate analogue to replace a phosphate residue in TAR RNA and shown covalent reaction with a specific lysine in Tat.

1 Introduction

The fascination of RNA is that it can take up a myriad of different three-dimensional structures and play widely varying roles within cells (Gesteland and Atkins, 1993). Despite being constructed of only four very similar nucleotide units, RNA has sufficient diversity in functional group chemistry to form many types of intra- and inter-molecular base pairs and it is also flexible enough to bind proteins as well as small molecule ligands. In addition, RNAs can carry out chemical reactions. Much effort has been expended on improving crystallographic and NMR techniques for RNA and these have at last provided the first structural models of biologically interesting RNAs and RNA–protein complexes. However, there is also an important need for a variety of complementary RNA techniques not only as tests of these early models but also in studies of the many RNAs that have not yet proved fully tractable to high resolution structural studies.

One complementary approach to aid studies of RNA structure and function involves the use of chemically synthesized model RNAs into which have been introduced site-specific modifications. Such modifications may include (i) individual functional group alterations or removals to test the requirements for RNA structural integrity or for RNA function (Usman and Cedergren, 1992; Grasby and Gait, 1994; Gait et al., 1995), (ii) the introduction of reactive groups that may allow chemical or photochemical RNA–RNA or RNA–protein cross-linking (Eaton and Pieken, 1995; Favre and Fourrey, 1995), or (iii) the introduction of specific labels or tags, for example fluorescent groups for resonance energy transfer experiments (Tuschl et al., 1994). We have been developing techniques involving the chemical synthesis of modified RNAs for studies of ribozymes and to aid in understanding of the interactions of HIV Tat and Rev proteins with their respective RNA recognition elements. These studies provide paradigms of the general utility of synthetic chemistry techniques towards RNA structure and function.

2 Chemical synthesis of oligoribonucleotides and analogues

Substantial improvements in oligoribonucleotide synthesis on solid supports were achieved in the early 1990s and led quickly to the commercial availability of routinely reliable RNA synthesis reagents. The predominant technique involves use of phosphoramidite coupling chemistry applied to nucleosides protected at the crucial 2'-hydroxyl position by the *t*-butyldimethylsilyl group (see Gait *et al.*, 1991; Damha and Ogilvie, 1993 for practical methodology). This 'silyl-phosphite' chemistry has been improved further in the last few years by the introduction of improved coupling agents, alternative solid-supports, base-protecting groups removable under milder conditions, faster desilylating conditions and higher resolution ion exchange HPLC columns (Gasparutto *et al.*, 1992; Sinha *et al.*, 1993; Reddy *et al.*, 1995; Sproat *et al.*, 1995; Tsou *et al.*, 1995; Wincott *et al.*, 1995). We have employed many of these improved methods (Schmidt *et al.*, 1996b; Gait *et al.*, 1997) and in our experience oligoribonucleotides up to about 40 residues may be assembled routinely on a small scale (1 μmole) to give 200 μg to 2 mg quantities. This scale synthesis is sufficient for most biochemical studies and three or four parallel syntheses provide enough material for crystallography (e.g. Scott *et al.*, 1995a, b). Larger scale syntheses have also been published (Davis, 1995; Tsou *et al.*, 1995; Wincott *et al.*, 1995).

In order to synthesize an oligoribonucleotide carrying a site-specific modification, it is necessary to prepare the modified ribonucleoside as a phosphoramidite derivative for solid-phase synthesis by the silyl-phosphite method. Despite a wide range of uses of modified ribonucleotides in studies of RNA structure and function (Usman and Cedergren, 1992; Grasby and Gait, 1994; Eaton and Pieken, 1995; Gait *et al.*, 1995), it is disappointing that very few modified ribonucleosides are currently available commercially as phosphoramidite derivatives.

2.1 Ribonucleotide analogues useful as RNA structural probes

There are a number of base analogues which have been found to be particularly useful for RNA structure–function analysis (Fig. 1). Inosine (hypoxanthine riboside) and nebularine (purine riboside) are analogues of guanosine and adenosine respectively where the exocyclic amino group is absent. The amino group is absolutely required for base-pairing of A and required in most types of base-pair involving G. Since also these modifications cause minimal disruption to the stacking properties of the base, they can be used to help prediction of base pairs within internal loops and bulges of RNAs. Inosine is also a good probe for structure-specific minor groove interactions of G amino groups in RNA duplex regions with proteins (Musier-Forsyth *et al.*, 1991; Iwai *et al.*, 1992; Hamy *et al.*, 1993). In addition to nebularine (Slim and Gait, 1992; Fu *et al.*, 1993), other useful purine nucleoside analogues are 2-aminoadenosine (Doudna *et al.*, 1990; SantaLucia Jr. *et al.*, 1991; Tuschl *et al.*, 1993), O^6-methylguanosine (Grasby *et al.*, 1993), N^7-deazaguanosine (Seela and Mersmann, 1992; Fu *et al.*, 1993;

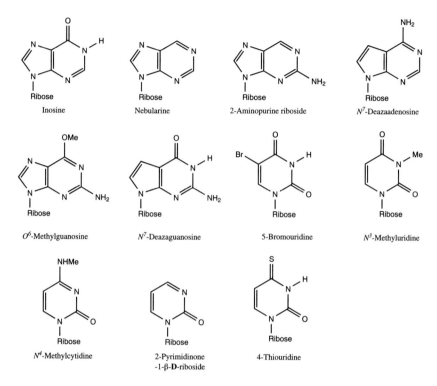

Fig. 1   Structures of the purine and pyrimidine base analogues used in RNA structure–function analysis.

Seela and Mersmann, 1993), and N^7-deazaadenosine (Fu and McLaughlin, 1992; Seela *et al.*, 1993). The N^7-deazanucleosides are valuable to probe the presence of Hoogsteen pairing interactions and chelation sites for metal ions as well as contacts to proteins.

Compared to purines, there are far fewer pyrimidine ribonucleoside analogues that have been incorporated chemically into oligoribonucleotides. 5-Bromouridine (5-BromoU) has proved valuable for X-ray crystallography for heavy atom derivatization and has been used for example in X-ray structure solution of a chemically synthesized hammerhead ribozyme (Scott *et al.*, 1995a, b). 5-BromoU was also the first ribonucleoside analogue to be incorporated synthetically into oligoribonucleotides for study of the interaction of MS2 protein with an RNA stem-loop (Talbot *et al.*, 1990). Other interesting pyrimidine analogues incorporated as phosphoramidites are N^3-methyluridine (Sumner-Smith *et al.*, 1991; Iwai *et al.*, 1992), N^4-methylcytidine (Grasby *et al.*, 1995b), and a fluorescent analogue, 2-pyrimidinone-1-β-D-riboside (Adams *et al.*, 1994a).

A more drastic modification of RNA is complete removal of a base. This is achieved by use of an abasic analogue (Fig. 2), which has been used in studies of

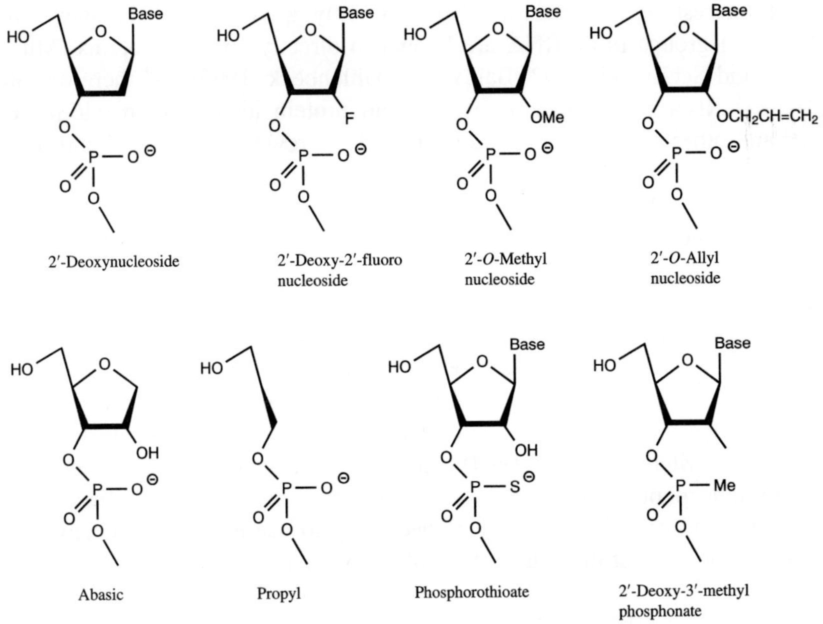

Fig. 2 Structures of sugar analogues and phosphate backbone modifications.

the hammerhead and hairpin ribozymes (Beigelman *et al.*, 1994; Schmidt *et al.*, 1996a). Another useful analogue (commercially available as an amidite) is a propyl linker. This is used to assess the need for both sugar and base together in a particular position in an RNA, since it maintains the number and type of atoms between neigbouring phosphate residues. For example, a propyl linker is tolerated in place of certain U residues in internal bulges in two cases of protein–RNA interactions (Sumner-Smith *et al.*, 1991; Iwai *et al.*, 1992) and in the hairpin ribozyme (Schmidt *et al.*, 1996a) (see below).

Individual substitutions in an RNA by 2′-deoxynucleosides have often been used as probes for the requirement of particular hydroxyl groups for ribozyme activity (Perreault *et al.*, 1990; Perreault *et al.*, 1991; Chowrira *et al.*, 1993). Single 2′-deoxynucleoside substitution in an RNA usually has little or no effect on the overall RNA conformation. By contrast, more drastic 2′-deoxynucleoside substitution, especially within RNA duplex regions, may affect the overall RNA conformation severely. This is because a 2′-deoxynucleoside usually adopts a 2′-*endo* (S) configuration, whereas most ribonucleosides in an RNA (except at bulged regions) usually adopt a 3′-*endo* (N) configuration. 2′-Deoxy-2′-fluoro, 2′-*O*-methyl or 2′-*O*-allyl nucleosides (Fig. 2) usually adopt 3′-*endo* configurations similar to that of a ribonucleoside. More global substitutions by such analogues are generally also tolerated, for example in synthetic ribozymes to improve nuclease resistance (Paolella *et al.*, 1992; Heidenreich *et al.*, 1994).

It is interesting to note that there are only a very few examples where 2′-hydroxyl groups in an RNA are known to directly contact proteins (Musier-Forsyth and Schimmel, 1992; Baidya and Uhlenbeck, 1995) and therefore often 2′-*O*-alkyl RNA probes can be used to aid protein identification within cells. These and other antisense applications have been reviewed (Lamond and Sproat, 1993). 2′-*O*-Allylnucleosides may also be incorporated into oligoribonucleotides (Sproat *et al.*, 1991). These have advantages of reduced non-specific protein binding in crude nuclear extracts when used as probes for RNA and are highly resistant to nucleases.

Two types of phosphate analogue may be incorporated into RNA with relative ease, a phosphorothioate and a 2′-deoxy-3′-methylphosphonate (Fig. 2). Both types of analogue are useful as probes of contacts with proteins (Milligan and Uhlenbeck, 1989; Pritchard *et al.*, 1994). The advantage of chemical synthesis is that the Rp and Sp isomers can usually be separated by HPLC (Slim and Gait, 1991; Pritchard *et al.*, 1994), and thus each can be assayed separately. A methylphosphonate linkage is unstable when incorporated adjacent to a 2′-hydroxyl group and therefore it is necessary to use a 2′-deoxynucleoside 3′-methylphosphonate at this site instead of a ribonucleoside.

2.2   Ribonucleotide analogues useful in RNA–RNA and RNA–protein cross-linking

Nucleosides containing thiolated bases, 4-thiouridine (4-thioU), 6-thioinosine, 6-thioguanosine and 2-thiouridine have been incorporated into synthetic oligoribonucleotides (Adams *et al.*, 1994b; Adams *et al.*, 1995a, b; Kumar and Davis, 1995). 4-thioU (Fig. 1) has had particularly wide usage for RNA–RNA cross-linking within ribozymes by irradiation of the RNA at 350–370 nm (reviewed in Favre and Fourrey, 1995). 4-thioU has also been used for some time for cross-linking ribosomal RNA to ribosomal proteins (Favre *et al.*, 1986) and now its incorporation into synthetic RNA allows cross-linking to other RNA binding proteins (McGregor *et al.*, 1996). 5-Bromouridine (Fig. 1) (Talbot *et al.*, 1990), and 5-iodouridine (Stump and Hall, 1995) are also useful photochemically activatable analogues for cross-linking to RNA-binding proteins.

One problem with photochemical cross-linking is that the chemistry can be complex, leading to the formation of multiple species. In addition it is difficult to control the sites of cross-linking. Promising new methods of chemical cross-linking within RNAs have recently been developed which obviate these problems. Intra- and interhelical disulfide cross-links via either two N^3-(thioethyl)uridines or two 2′-*O*-alkylthio modified cytosine residues have been introduced into particular sites in yeast tRNA^{Phe} without perturbation (Goodwin *et al.*, 1996).

An alternative disulfide cross-linking route involves introduction of 2′-amino-2′-deoxynucleosides into synthetic RNAs (Pieken *et al.*, 1991). Reaction of two specifically placed 2′-amino groups in a synthetic RNA with an aryl isothiocyanate derivative containing a 2-pyridyl disulfide followed by reduction and

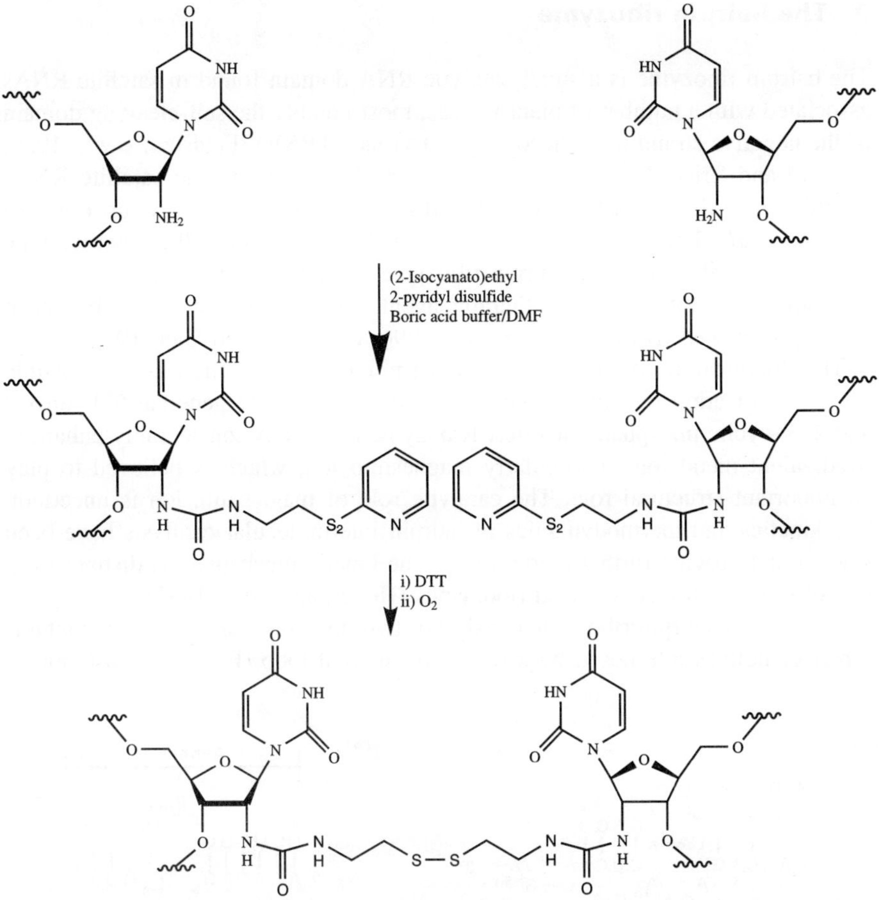

Fig. 3 Disulfide cross-linking of 2′-positions of nucleosides in RNA. 2′-amino-2′-deoxy uridines are reacted with an alkyl isocyanate reagent, pyridyl groups removed, and oxidized to give a disulfide cross-link.

oxidation allows disulfide joining of the two selected nucleoside 2′-positions (Sigurdsson *et al.*, 1995). A more flexible but analogous alkylisocyanate derivative can also be used to label 2′-positions specifically in an oligoribonucleotide (Sigurdsson and Eckstein, 1996) or for cross-linking (Fig. 3). We have used disulfide cross-linking methods in studies of the hairpin ribozyme (see below).

We have also developed recently a new procedure for specific reaction of lysine residues in a protein with a closely located phosphate residue in an RNA recognition element (Naryshkin *et al.*, 1997). The method involves the introduction, *via* a template-dependent chemical ligation reaction, of a trisubstituted pyrophosphate (tsp) analogue in place of a particular phosphate residue in a model TAR RNA duplex. The tsp analogue was shown to react covalently with a specific lysine residue in an HIV-1 Tat peptide (see below).

3 The hairpin ribozyme

The hairpin ribozyme is a small catalytic RNA domain found in satellite RNAs associated with a number of plant viruses, most notably the self-cleaving domain of the negative strand of tobacco ringspot virus (sTRSV) (Feldstein *et al*., 1989; Hampel and Tritz, 1989; Haseloff and Gerlach, 1989) and the satellite RNAs of chicory yellow mottle virus (sCYMV) and arabis mosaic virus (sAMV) (Rubino *et al*., 1990; DeYoung *et al*., 1995). Although less well known than the hammerhead ribozyme, the hairpin ribozyme has already attracted considerable attention as a potential gene therapy agent. The hairpin ribozyme has been recently reviewed (Burke, 1994; Burke, 1996; Earnshaw and Gait, 1997).

The chemical reaction involved in hairpin ribozyme action is a reversible *trans*-esterification resulting in phosphodiester cleavage to generate 5′-hydroxyl and 2′,3′-cyclic phosphate moieties. Ribozyme activity is considerably enhanced by divalent metal ions, particularly magnesium ion, which is believed to play an important structural role. The catalytic role of magnesium ion is uncertain. The kinetics and thermodynamics of hairpin intermolecular catalysis have been studied and provide further evidence that the kinetic mechanism is distinct from that of other well-characterized ribozymes (Hegg and Fedor, 1995).

The sTRSV hairpin ribozyme consists of two domains, each of which contains a pair of helices interposed by a region of internal loop (Fig. 4a). Most studies

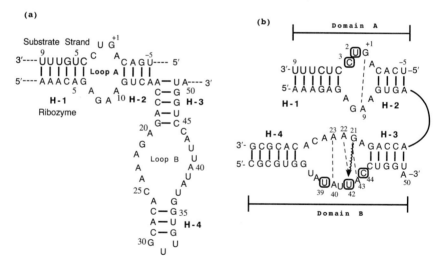

Fig. 4 (a) Secondary structure of the (sTRSV) hairpin ribozyme. H = helix; (b) A three stranded hairpin ribozyme with the terminal loop of helix 4 removed. Folding of the two domains allows interaction between the nucleotides within the two loops. Proposed cross-strand sheared G:A base pair in loop A (dashed line) is based on NMR data (Cai and Tinoco, 1996). Loop B non-canonical base pairs (dashed lines) are as proposed (Butcher and Burke, 1994b). The curly arrow denotes a UV cross-link (Butcher and Burke, 1994a). Boxed residues indicate incorporation of 2′-amino-2′-deoxynucleosides for cross-linking studies.

have concentrated on *trans*-cleaving hairpin ribozymes, especially towards the possibility of therapeutic applications. Control of the number of base-pairs between substrate and ribozyme is important. For example, if the number of base pairs in helix 1 is increased, rates of cleavage *in vitro* are limited by slow product dissociation (Hegg and Fedor, 1995). Fortunately, for substrates containing six pairs in helix 1 and four pairs in helix 2, dissociation rates are faster than cleavage. For such hairpins K_M is in the range 20–80 nM and k_{cat} about 0.1 to 0.3 min^{-1}, similar to the k_{cat} values for other ribozymes.

The binding rate constant of the *trans*-cleaving sTRSV ribozyme is limited by an alternative conformation of the substrate. Two base-pair switches in the flanks improve the binding 50-fold to about 3×10^8 M^{-1} min^{-1} (Esteban *et al.*, 1997), which is amongst the fastest observed between nucleic acids. There is also evidence for biphasic kinetics of ribozyme cleavage, thought due to two non-exchangeable conformations, one being inactive, where helixes 2 and 3 are coaxially stacked, the other active form being correctly folded and resulting in the fast cleavage phase (Esteban *et al.*, 1997).

3.1 Structure–function analysis of the hairpin ribozyme

In the complete sTRSV RNA, the hairpin is part of a larger structure and it is proposed that the junction of domains A and B is four-way, which may allow alternative co-axial stacking or folding conformers (Duckett *et al.*, 1995). The active catalytic conformer is likely to be one where the two domains approach each other closely by bending *via* a hinge at the junction of helixes 2 and 3 (Fig. 4b) (Feldstein and Bruening, 1993; Komatsu *et al.*, 1994). Further, the two domains can be completely separated into independent sections and ribozyme activity reconstituted by addition of one domain to the other (Butcher *et al.*, 1995; Komatsu *et al.*, 1996; Shin *et al.*, 1996).

So far there are no crystallographic or NMR structures for the hairpin ribozyme. However, a solution structure for isolated domain A has been proposed based on NMR spectroscopic measurements. This provided evidence for a number of cross-strand interactions in loop A including a sheared base pair between G_{+1} and A_9 (Cai and Tinoco, 1996). The addition of magnesium ion had little effect on the domain A folded structure. Domain B folds independently of substrate binding, but there are some major differences in chemical accessibilities of residues in loop B when magnesium ion is added (Butcher and Burke, 1994b). A proposed secondary structure for loop B that includes three non-canonical base pairs between $G_{21}:A_{43}$, $A_{22}:U_{42}$ and $A_{23}:A_{40}$ (Fig. 4b) is based on chemical accessibility data as well as the finding that a stable UV cross-link can be formed between G_{21} and U_{42} (Butcher and Burke, 1994a; Burke, 1996), similarly to other UV sensitive domains such as those found in the loop E of eukaryotic 5S RNA and in the conserved C domain of viroids.

In vitro selection and mutagenesis studies have been applied to determine the residues essential for cleavage of the hairpin ribozyme (Berzal-Herranz *et al.*,

1992; Berzal-Herranz *et al.*, 1993; Joseph *et al.*, 1993; Joseph and Burke, 1993; Anderson *et al.*, 1994). Apart from the maintenance of the four helixes, there is little requirement for sequence identity in the helical regions. By contrast, there are strict requirements for many of the nucleotides in the two internal loops. In the substrate strand, only the G_{+1} 3′ to the cleaved phosphodiester is uniquely defined (Joseph *et al.*, 1993) and this in principle allows a wide range of possible RNA sites to be addressed for therapeutic ribozyme design.

We and others have carried out more detailed mapping of the functional group requirements for ribozyme cleavage of the essential residues in loops A and B. Site-specific modifications were introduced *via* chemical synthesis, mostly using a three-stranded ribozyme where helix 4 is extended by 3 base pairs and the terminal loop is omitted (Chowrira *et al.*, 1993; Grasby *et al.*, 1995a). In the substrate strand, removal of the exocyclic amino group at G_{+1} (substitution by inosine) resulted in drastic loss of ribozyme activity (at least three orders) (Chowrira and Burke, 1991; Grasby *et al.*, 1995a). Removal of exocyclic amino groups or substitution by N^7-deazapurine nucleosides at any of positions A_7, A_8, G_9 or A_{10} in loop A resulted in significant loss of ribozyme activity (Grasby *et al.*, 1995a), mostly reflected in changes in k_{cat} values, and were consistent with loss of hydrogen-bonding contacts either across loop A or inter-domain.

In loop B there were significant effects on k_{cat} values upon removal of exocyclic amino groups or N^7-deazapurine nucleoside substitution at either G_{21}, A_{22} or A_{23}, but there was hardly any effect of removal of the exocyclic amino group from either A_{40} or A_{43} (Grasby *et al.*, 1995a). These results suggest that some of the predicted base pairs across loop B in the ground state may not be present in the transition state, since all possible base pairs involving an A residue would require the exocyclic amino group.

More recently we have carried out a detailed substitution study of residues in loop B by either a 2′-deoxynucleoside, an abasic residue or a propyl linker (Schmidt *et al.*, 1996a). We found that substitution of U_{39} by an abasic residue or by a propyl linker had no effect on ribozyme cleavage and that this nucleoside is likely therefore to be acting merely as a spacer. Similarly the residue at C_{44} can be replaced by an abasic residue without significant loss of catalytic activity. Other results were consistent with the important roles of many of the other residues in loop B.

Replacement of any of four hydroxyl groups in the ribozyme strand by 2′-deoxy or 2′-*O*-methyl at A_{10}, G_{11}, A_{24} or C_{25} was found to be detrimental to ribozyme activity (Chowrira *et al.*, 1993). This inhibition could be rescued for two of these sites (G_{11} or A_{24}) by an increase in the magnesium ion concentration, suggesting that these sites may be important in the active site architecture of the transition state. A possible ground state contact with magnesium at the N^7-position of G_{+1} is indicated by the finding that N^7-deaza-G substituted mutant ribozyme had a significantly increased apparent magnesium binding constant (Grasby *et al.*, 1995a). Loss of ribozyme activity by N^7-deazaadenosine substitution at A_9 could be partially restored by an increase in magnesium ion concentration, suggesting a possible contact of magnesium ion with the N^7 position of A_9 in the transition state.

3.2 Folding of the hairpin ribozyme

It is now reasonably well established that domain B folds independently of substrate binding (formation of domain A). There is also much evidence that the two domains must interact closely in the transition state. Recent studies suggest that 2.1 kcal mole^{-1} of additional substrate binding energy is provided by the presence of domain B and at least part of this energy must be due to inter-domain interactions (Walter and Burke, 1997).

We have been developing a chemical cross-linking approach to help understand how the two domains interact (Earnshaw *et al.*, 1997). Specific pyrimidine nucleoside residues in both substrate strand (loop A) and ribozyme strand (loop B) were replaced by 2'-amino-2'-deoxynucleosides and reacted with either an aryl isothiocyanate (Sigurdsson *et al.*, 1995) or an alkyl isocyanate (Sigurdsson and Eckstein, 1996), each containing a pyridyl disulfide. Upon reduction to remove pyridyl groups followed by oxidation, intermolecular disulfide bonds were formed. Cross-linked substrate-ribozyme strands B were each annealed to ribozyme strand A to form inter-domain constrained hairpin ribozymes. A comparison of the single turnover kinetic parameters of the cross-linked with their disulfide-reduced and hence unconstrained counterparts has identified three classes of tethers: (a) those which had little or no effect on ribozyme cleavage (e.g. U_2-U_{39}), (b) those where there was a moderate effect (e.g. C_3-U_{39}) and (c) those where there was a drastic loss of activity upon cross-linking (e.g. C_3-C_{44}). An initial folding model for the hairpin ribozyme has been constructed and the docking of the two domains was guided by the rank order of ribozyme activity for cross-links from U_2 and C_3 in loop A to each of three positions in loop B. The folding model of the hairpin ribozyme can now act as a basis for the design of further cross-linking or other structure-testing experiments.

4 HIV-1 Tat protein interaction with TAR RNA

The Human Immunodeficiency Virus type 1 (HIV-1) *trans*-activator protein Tat is essential to viral replication (for a recent review see Karn *et al.*, 1996). In the absence of Tat, HIV transcripts are prematurely terminated. As Tat levels rise during gene expression, a considerable boost is obtained in the amount of full-length transcripts. Activity of Tat is triggered by recognition of an RNA sequence, the *trans*-activation response region TAR, which occurs at the 5'-end of all viral RNAs. TAR is a 59-residue RNA stem-loop which binds to Tat *in vitro* with high affinity (Dingwall *et al.*, 1989; Dingwall *et al.*, 1990; Müller *et al.*, 1990; Roy *et al.*, 1990). Recognition of this TAR RNA is dependent on a section of the HIV-1 Tat protein consisting of a basic region which is particularly rich in arginine, flanked on one side by a highly conserved hydrophobic core region and on the other by a section of more variable sequence but containing several glutamine residues (Fig. 5a). The flanking core and glutamine-rich regions influence the overall folding of Tat and help to orient the basic region for interaction with TAR (Bayer *et al.*, 1995).

Because of difficulties in obtaining correctly folded full length Tat, many Tat–TAR binding studies have focused on the use of Tat peptides. Peptides spanning the basic region alone (Cordingley *et al.*, 1990; Calnan *et al.*, 1991a), or the basic and glutamine-rich regions (Weeks *et al.*, 1990; Weeks and Crothers, 1991; Long and Crothers, 1995), can bind TAR with high affinity. Even a single arginine derivative (argininamide) can bind to TAR with K_d in the millimolar range (cf. Tat -TAR K_d in the nM range) (Tao and Frankel, 1992) and give rise to structural changes in TAR (Tan and Frankel, 1992) similar to that elicited by longer peptides. However, to obtain the full specificity of the Tat–TAR interaction, a larger region of Tat is required (e.g. residues 37–72, Fig. 5a), that includes the hydrophobic core and glutamine-rich regions (Kamine *et al.*, 1991; Churcher *et al.*, 1993).

(a)

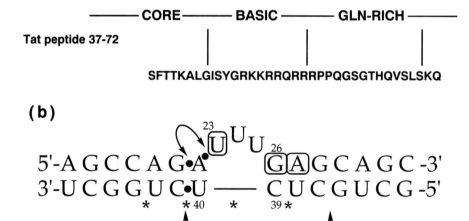

(b)

Fig. 5 (a) Regions and sequence of HIV-1 Tat peptide 37–72; (b) Structure of HIV-1 TAR synthetic model duplex containing a U-rich bulge to which Tat binds. Stars denote sites of incorporation of trisubstituted pyrophosphates. Boxes indicate bases whose functional groups are crucial to Tat binding. Arrows denote the positions of ethylation protection observed (Calnan *et al.*, 1991b; Hamy *et al.*, 1993). Filled circles denote phosphates, substitution of which by a methylphosphonate (R_p or S_p) leads to inhibition of Tat binding.

4.1 Functional Group mapping of the Tat–TAR interaction

The region of TAR recognized by Tat is near the apex and contains a 3-residue U-rich bulge (Dingwall *et al*., 1990; Roy *et al*., 1990). Mutation of one of these residues, U$_{23}$, or either of the two base pairs above the bulge leads to a drastic reduction in Tat binding affinity (Delling *et al*., 1992; Churcher *et al*., 1993). The apical loop is not required for Tat binding and can be dispensed with in TAR models (Sumner-Smith *et al*., 1991; Hamy *et al*., 1993).

We and others have been able to identify a number of essential functional groups on base residues by chemical synthesis of model TAR duplexes (e.g. Fig. 5b) containing singly modified base residues and by comparison of their Tat binding abilities with unmodified TAR duplex. Critical to the interaction are the N^7 positions of G$_{26}$ and A$_{27}$ in the two base pairs above the bulge and the O^4 and N^3-H positions of U$_{23}$ in the bulge (Sumner-Smith *et al*., 1991; Hamy *et al*., 1993), all of which would be expected to be oriented towards the major groove of an A-form RNA duplex structure. Phosphate residues are also important to Tat binding. Ethylation interference analysis showed two regions of TAR protection, two phosphates on one strand (P$_{21-22}$ and P$_{22-23}$) and five phosphates on the other strand (between residues 36 and 41) (Fig. 5b) (Calnan *et al*., 1991b; Hamy *et al*., 1993). We carried out methylphosphonate substitution analysis using chemically synthesized TAR analogues (where an individual oxygen atom (R_p or S_p) on a phosphate is replaced by a methyl group) and identified three phosphates (P$_{21-22}$, P$_{22-23}$ and P$_{40-41}$) where methylphosphonate substitution in each case is severely harmful to Tat binding (Pritchard *et al*., 1994). These phosphates are likely candidates for direct interactions with Tat, possibly with basic residues such as arginine or lysine.

4.2 The NMR structure of TAR

Recent NMR studies of free TAR (Aboul-ela *et al*., 1996) and in the presence of Tat peptide 37–72 (Aboul-ela *et al*., 1995) showed that the major groove of TAR in the region of the U-rich bulge is wider than for normal A-type RNA helixes, but is narrowed considerably upon peptide binding. Both the Tat peptide (Aboul-ela *et al*., 1995) and argininamide (Puglisi *et al*., 1992; Puglisi *et al*., 1993; Aboul-ela *et al*., 1995) induce a similar conformational change suggesting that one key arginine residue in the basic region of Tat is sufficient to trigger this, possibly by interaction with G$_{26}$ and U$_{23}$. In the bound form of TAR, all the important functional groups identified by chemical substitution experiments become clustered close to the surface of the duplex on the major groove side and are available for interaction with Tat. The identified phosphate residues may make essential contributions to the Tat–TAR recognition process via 'indirect readout' through their precise positioning away from locations normally found for regular RNA duplexes (Aboul-ela *et al*., 1996).

4.3 Cross-linking of TAR RNA duplex to HIV-1 Tat peptides

An alternative method capable of providing valuable information about possible contacts of specific amino acids to particular RNA residues is cross-linking. For example, an N-terminal psoralen conjugate of Tat peptide 42–72 was able to bind specifically to TAR and the psoralen photochemically cross-linked to U_{42} in the lower stem (Wang and Rana, 1995). Another psoralen conjugate of Tat peptide 38–72 to a unique Cys[57] residue, introduced synthetically to replace Arg[57], cross-linked to the TAR apical loop at position U_{31} (Wang et al., 1996). Further, Tyr[47] in the core region of Tat peptide 42–72 could be specifically UV-cross-linked to G_{26} (Liu et al., 1996). These experiments have suggested that the orientation of the basic region of Tat lies from N to C in the direction from lower TAR stem to upper TAR stem and loop.

We have been interested in a second cross-linking strategy where reactive cross-linking functionalities are introduced on TAR which are then reacted with a Tat peptide in order to identify acceptor sites. It has been reported that a TAR duplex containing 4-thioU at position 23 can be photochemically cross-linked to Tat peptide 38–72, but the position(s) of cross-linking on the peptide could not be determined (Wang and Rana, 1996). We have confirmed this cross-linking result and localized the cross-linking to the basic region of Tat peptide 38–72 (Farrow, M. and Gait, M. J., unpublished results).

We have been interested in determining which residues in Tat might approach closely to particular phosphates in TAR. This could prove very useful since NMR studies of protein–RNA complexes are currently only able to give indirect information about such contacts. Recently, a new method has been described for the site-specific incorporation of an activated trisubstituted pyrophosphate (tsp) into synthetic DNA duplexes (Kuznetsova et al., 1990; Purmal et al., 1992; Kuznetsova et al., 1996). It was shown that only minor distortion of the double helix occurs upon incorporation of this analogue. A DNA containing the recognition sequence of a restriction nuclease and substituted at the scissile bond by a tsp was found to be reactive towards active site nucleophiles to form a covalent phosphoamidate bond when incubated with a restriction enzyme (Purmal et al., 1992). Formation of the cross-link was limited to specific DNA–enzyme recognition.

We have successfully incorporated tsps into several sites in TAR RNA model duplexes and shown that tsp linkages are reasonably stable to hydrolysis but react rapidly with ethylene diamine or lysine, but not with N-ethylguanidine (Naryshkin et al., 1997). TAR RNA duplexes carrying tsp modifications at phosphates expected to be in close proximity to Tat reacted specifically and covalently with Tat peptide 37–72 or with Tat protein. Analysis of the cross-linked product of TAR containing a tsp at site P_{38-39} in the upper stem with Tat peptide 37–72 indicated reaction exclusively with the epsilon amino group of Lys[51] in the basic region of Tat (Naryshkin et al., 1997). The result is consistent with a preliminary model of the Tat–TAR complex based on NMR data where the basic region

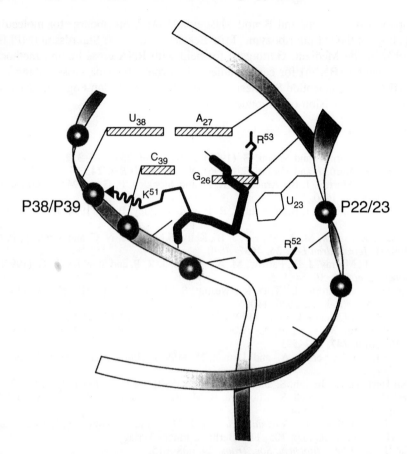

Fig. 6 Schematic representation of the model of the TAR RNA structure in the region of the U-rich bulge with a section of bound Tat peptide (Arg^{53}-Arg^{52}-Lys^{51}) showing how Lys^{51} is favourably placed for cross-linking to P_{38-39}. The important base-pairs immediately above the bulge ($G_{26} \cdot C_{39}$ and $A_{27} \cdot U_{38}$) are shown hatched and are viewed edgewise whereas the crucial U_{23} is in a different plane.

of the Tat peptide is inserted into the major groove of TAR (Aboul-ela, F. and Varani, G., personal communication). There are proposed contacts of Arg^{53} with G_{26} and A_{27}, Arg^{52} with P_{21-22} and P_{22-23}, such that Lys^{51} is in a position to form a cross-link easily with the opposite strand phosphates including P_{38-39} (Fig. 6). This model provides a basis for further chemical cross-linking experiments in progress.

Acknowledgements

We thank our collaborators Fareed Aboul-ela and Gabriele Varani (MRC, Cambridge) for their work on the NMR structure of HIV-1 TAR RNA–Tat peptide complex, Eric Westhof and Benoît Masquida (IBMC, Strasbourg) for molecular modelling of the hairpin ribozyme, Fritz Eckstein and Snorri Sigurdsson (MPI für Experimentelle Medizin, Göttingen) for help with RNA cross-linking methods, Sabine Müller (Berlin) for early studies on hairpin ribozyme cross-linking and the MRC Oligonucleotide Synthesis Service (Terry Smith, Jan Fogg and Richard Grenfell) for provision of oligonucleotides.

References

Aboul-ela, F., Karn, J. and Varani, G. (1995). *J. Mol. Biol.* **253**, 313–332.
Aboul-ela, F., Karn, J. and Varani, G. (1996). *Nucleic Acids Res.* **24**, 3974–3982.
Adams, C. J., Murray, J. B., Arnold, J. R. P. and Stockley, P. G. (1994a). *Tetrahedron Lett.* **35**, 1597–1600.
Adams, C. J., Murray, J. B., Arnold, J. R. P. and Stockley, P. G. (1994b). *Tetrahedron Lett.* **35**, 765–768.
Adams, C. J., Farrow, M. A., Murray, J. B., Kelly, S.M., Price, N. C. and Stockley, P. G. (1995a). *Tetrahedron Lett.* **36**, 4637–4640.
Adams, C. J., Murray, J. B., Farrow, M. A., Arnold, J. R. P. and Stockley, P. G. (1995b). *Tetrahedron Lett.* **36**, 5421–5424.
Anderson, P., Monforte, J., Tritz, R., Nesbitt, S., Hearst, J. and Hampel, A. (1994). *Nucleic Acids Res.* **22**, 1096–1100.
Baidya, N. and Uhlenbeck, O. C. (1995). *Biochemistry* **34**, 12363–12368.
Bayer, P., Kraft, M., Ejchart, A., Westendorp, M., Frank, R. and Rösch, P. (1995). *J. Mol. Biol.* **247**, 529–535.
Beigelman, L., Karpeisky, A. and Usman, N. (1994). *Biorg. Med. Chem. Letters* **4**, 1715–1720.
Berzal-Herranz, A., Joseph, S. and Burke, J. M. (1992). *Genes and Dev.* **6**, 129–134.
Berzal-Herranz, A., Joseph, S., Chowrira, B. M., Butcher, S. E. and Burke, J. M. (1993). *EMBO J.* **12**, 2567–2574.
Burke, J. M. (1994). In *Nucleic Acids and Molecular Biology* (F. Eckstein and D. M. J. Lilley, eds), pp. 105–118. Berlin, Springer-Verlag.
Burke, J. M. (1996). *Biochem. Soc. Trans.* **24**, 608–615.
Butcher, S. E. and Burke, J. M. (1994a). *Biochemistry* **33**, 992–999.
Butcher, S. E. and Burke, J. M. (1994b). *J. Mol. Biol.* **244**, 52–63.
Butcher, S. E., Heckman, J. E. and Burke, J. M. (1995). *J. Biol. Chem.* **270**, 29648–29652.
Cai, Z. and Tinoco, I. (1996). *Biochemistry* **35**, 6026–6036.
Calnan, B. J., Biancalana, S., Hudson, D. and Frankel, A. D. (1991a). *Genes & Develop.* **5**, 201–210.
Calnan, B. J., Tidor, B., Biancalana, S., Hudson, D. and Frankel, A. D. (1991b). *Science* **252**, 1167–1171.
Chowrira, B. M. and Burke, J. M. (1991). *Nature* **354**, 320–322.
Chowrira, B. M., Berzal-Herranz, A., Keller, C. F. and Burke, J. M. (1993). *J. Biol. Chem.* **268**, 19458–19462.
Churcher, M. J., Lamont, C., Hamy, F., Dingwall, C., Green, S. M., Lowe, A. D., Butler, P. J. G., Gait, M. J. and Karn, J. (1993). *J. Mol. Biol.* **230**, 90–110.
Cordingley, M. G., LaFemina, R. L., Callahan, P. L., Condra, J. H., Sardana, V. V., Graham, D. J., Nguyen, T. M., LeGrow, K., Gotlib, L., Schlabach, A. J. and Colonno, R. J. (1990). *Proc. Natl. Acad. Sci. USA* **87**, 8985–8989.
Damha, M. J. and Ogilvie, K. K. (1993). In *Methods in Molecular Biology* (S. Agrawal and N. J. Totowa, eds), pp. 81–114. Humana Press.
Davis, R. H. (1995). *Curr. Opin. in Biotech.* **6**, 213–217.

Delling, U., Reid, L. S., Barnett, R. W., Ma, M. X. Y., Climie, S., Sumner-Smith, M. and Sonenberg, N. (1992). *J. Virol.* **66**, 3018–3025.
DeYoung, M. B., Siwkowski, A. M., Lian, Y. and Hampel, A. (1995). *Biochemistry* **34**, 15785–15791.
Dingwall, C., Ernberg, I., Gait, M. J., Green, S. M., Heaphy, S., Karn, J., Lowe, A. D., Singh, M., Skinner, M. A. and Valerio, R. (1989). *Proc. Natl. Acad. Sci. USA* **86**, 6925–6929.
Dingwall, C., Ernberg, I., Gait, M. J., Green, S. M., Heaphy, S., Karn, J., Lowe, A. D., Singh, M. and Skinner, M. A. (1990). *EMBO J.* **9**, 4145–4153.
Doudna, J. A., Szostak, J. W., Rich, A. and Usman, N. (1990). *J. Org. Chem.* **55**, 5547–5549.
Duckett, R., Murchie, A. I. H. and Lilley, D. M. J. (1995). *Cell* **83**, 1027–1036.
Earnshaw, D. J. and Gait, M. J. (1997). *Antisense & Nucl. Acid Drug Dev.* **7**, 405–413.
Earnshaw, D. J., Masquida, B., Müller, S., Sigurdsson, S., Eckstein, F., Westhof, E. and Gait, M. J. (1997). *J. Mol. Biol.*, in press.
Eaton, B. E. and Pieken, W. A. (1995). *Annu. Rev. Biochem.* **64**, 837–863.
Esteban, J. A., Banerjee, A. R. and Burke, J. M. (1997). *J. Biol. Chem.* **272**, 13629–13639.
Favre, A., Bezerra, R., Hajnsdorf, E., Dubreuil, L. and Expert-Bezançon, A. (1986). *Eur. J. Biochem.* **160**, 441–449.
Favre, A. and Fourrey, A. (1995). *Acc. Chem. Res.* **28**, 375–382.
Feldstein, P. A. and Bruening, G. (1993). *Nucl. Acids Res.* **21**, 1991–1998.
Feldstein, P. A., Buzayan, J. M. and Bruening, G. (1989). *Gene* **82**, 53–61.
Fu, D.-J. and McLaughlin, L. W. (1992). *Biochemistry* **31**, 10941–10949.
Fu, D.-J., Rajur, S. B. and McLaughlin, L. W. (1993). *Biochemistry* **32**, 10629–10637.
Gait, M. J., Pritchard, C. E. and Slim, G. (1991). In *Oligonucleotides and Analogues: A Practical Approach* (F. Eckstein, ed.), pp. 25–48. Oxford, UK, Oxford University Press.
Gait, M. J., Grasby, J., Karn, J., Mersmann, K. and Pritchard, C. E. (1995). *Nucleosides and Nucleotides* **14**, 1133–1144.
Gait, M. J., Earnshaw, D. J., Farrow, M. A., Fogg, J. H., Grenfell, R. L., Naryshkin, N. A. and Smith, T. V. (1997). In *RNA–protein Interactions: A Practical Approach.* (C. Smith, ed.), in press. Oxford, UK, Oxford University Press.
Gasparutto, D., Livache, T., Bazin, H., Duplaa, A.-M., Guy, A., Khorlin, A., Molko, D., Roget, A. and Teoule, R. (1992). *Nucl. Acids Res.* **19**, 5159–5166.
Gesteland, R. F. and Atkins, J. F. (1993). *The RNA World.* New York, Cold Spring Harbor Press.
Goodwin, J. T., Osborne, S. E., Scholle, E. J. and Glick, G. D. (1996). *J. Am. Chem. Soc.* **118**, 5207–5215.
Grasby, J. A. and Gait, M. J. (1994). *Biochimie* **76**, 1223–1234.
Grasby, J. A., Butler, P. J. G. and Gait, M. J. (1993). *Nucleic Acids Res.* **21**, 4444–4450.
Grasby, J., Mersmann, K., Singh, M. and Gait, M. J. (1995a). *Biochemistry* **34**, 4068–4076.
Grasby, J. A., Singh, M., Karn, J. and Gait, M. J. (1995b). *Nucleosides and Nucleotides* **14**, 1129–1132.
Hampel, A. and Tritz, R. (1989). *Biochemistry* **28**, 4929–4933.
Hamy, F., Asseline, U., Grasby, J. A., Iwai, S., Pritchard, C. E., Slim, G., Butler, P. J. G., Karn, J. and Gait, M. J. (1993). *J. Mol. Biol.* **230**, 111–123.
Haseloff, J. and Gerlach, W. L. (1989). *Gene* **82**, 43–52.
Hegg, L. A. and Fedor, M. J. (1995). *Biochemistry* **34**, 15813–15828.
Heidenreich, O., Benseler, F., Fahrenholz, A. and Eckstein, F. (1994). *J. Biol. Chem.* **269**, 2131–2138.
Iwai, S., Pritchard, C. E., Mann, D. A., Karn, J. and Gait, M. J. (1992). *Nucleic Acids Res.* **20**, 6465–6472.

Joseph, S., Berzal-Herranz, A., Chowrira, B. M., Butcher, S. E. and Burke, J. M. (1993). *Genes and Dev.* **7**, 130–138.

Joseph, S. and Burke, S. (1993). *J. Biol. Chem.* **268**, 24515–24518.

Kamine, J., Loewenstein, P. and Green, M. (1991). *Virology* **182**, 570–577.

Karn, J., Churcher, M. J., Rittner, K., Keen, N. and Gait, M. J. (1996). In *Eukaryotic Gene Transcription* (S. Goodbourn, ed.), pp. 254–286. Oxford, Oxford University Press.

Komatsu, Y., Koizumi, M., Nakamura, H. and Ohtsuka, E. (1994). *J. Am. Chem. Soc.* **116**, 3692–3696.

Komatsu, Y., Kanzaki, I. and Ohtsuka, E. (1996). *Biochemistry* **35**, 9815–9820.

Kumar, R. K. and Davis, D. R. (1995). *J. Org. Chem.* **60**, 7726–7727.

Kuznetsova, S. A., Ivanovskaya, M. G. and Shabarova, Z. A. (1990). *Bioorg. Khim.* **16**, 219–225.

Kuznetsova, S. A., Blumenfeld, M., Vasseur, M. and Shabarova, Z. A. (1996). *Nucleosides and Nucleotides* **15**, 1237–1251.

Lamond, A. I. and Sproat, B. S. (1993). *FEBS Letters* **325**, 123–127.

Liu, Y., Wang, Z. and Rana, T. M. (1996). *J. Biol. Chem.* **271**, 10391–10396.

Long, K. S. and Crothers, D. M. (1995). *Biochemistry* **34**, 8885–8895.

McGregor, A., Rao, M. V., Duckworth, G., Stockley, P. G. and Connolly, B. A. (1996). *Nucleic Acids Res.* **24**, 3173–3180.

Milligan, J. F. and Uhlenbeck, O. C. (1989). *Biochemistry* **28**, 2849–2855.

Müller, W. E. G., Okamoto, T., Reuter, P., Ugarkovic, D. and Schröder, H. C. (1990). *J. Biol. Chem.* **265**, 3803–3808.

Musier-Forsyth, K. and Schimmel, P. (1992). *Nature* **357**, 513–515.

Musier-Forsyth, K., Usman, N., Scaringe, S., Doudna, J., Green, R. and Schimmel, P. (1991). *Science* **253**, 784–786.

Naryshkin, N. A., Farrow, M. A., Ivanovskaya, M. G., Orestkaya, T. S., Shabarova, Z. A. and Gait, M. J. (1997). *Biochemistry* **36**, 3496–3505.

Paolella, G., Sproat, B. and Lamond, A. I. (1992). *EMBO J.* **11**, 1913–1919.

Perreault, J.-P., Wu, T., Cousineau, B., Ogilvie, K. K. and Cedergren, R. (1990). *Nature* **344**, 565–567.

Perreault, J.-P., Labuda, D., Usman, N., Yang, J.-H. and Cedergren, R. (1991). *Biochemistry* **30**, 4020–4025.

Pieken, W. A., Olsen, D. B., Benseler, F., Aurup, H. and Eckstein, F. (1991). *Science* **253**, 314–317.

Pritchard, C. E., Grasby, J. A., Hamy, F., Zacharek, A. M., Singh, M., Karn, J. and Gait, M. J. (1994). *Nucleic Acids Res.* **22**, 2592–2600.

Puglisi, J. D., Tan, R., Calnan, B. J., Frankel, A. D. and Williamson, J. R. (1992). *Science* **257**, 76–80.

Puglisi, J. D., Chen, L., Frankel, A. D. and Williamson, J. R. (1993). *Proc. Natl. Acad. Sci. USA* **90**, 3680–3684.

Purmal, A. A., Shabarova, Z. A. and Gumport, R. I. (1992). *Nucleic Acids Res.* **20**, 3713–3719.

Reddy, M. P., Farooqui, F. and Hanna, N. B. (1995). *Tetrahedron Letts.* **36**, 8929–8932.

Roy, S., Delling, U., Chen, C.-H., Rosen, C. A. and Sonenberg, N. (1990). *Genes & Devel.* **4**, 1365–1373.

Rubino, L., Tousignant, M. E., Steger, G. and Kaper, J. M. (1990). *J. Gen. Virol.* **71**, 1897–1903.

SantaLucia Jr., J., Kierzek, R. and Turner, D. H. (1991). *J. Am. Chem. Soc.* **113**, 4313–4322.

Schmidt, S., Beigelman, L., Karpeisky, A., Usman, N., Sørensen, U. S. and Gait, M. J. (1996a). *Nucleic Acids Res.* **24**, 573–581.

Schmidt, S., Grenfell, R. L., Fogg, J., Smith, T. V., Grasby, J. A., Mersmann, K. and Gait, M. J. (1996b). In *Innovation and Perspectives in Solid Phase Synthesis and Combinatorial Libraries, 4th International Symposium Proceedings 1996* (R. Epton, ed.), pp. 11-18. Mayflower, Birmingham, UK.

Scott, W. G., Finch, J. T. and Klug, A. (1995a). *Cell* **81**, 991-1002.

Scott, W. G., Finch, J. T., Grenfell, R., Fogg, J., Smith, T., Gait, M. J. and Klug, A. (1995b). *J. Mol. Biol.* **250**, 327-332.

Seela, F. and Mersmann, K. (1992). *Heterocycles* **34**, 229-236.

Seela, F. and Mersmann, K. (1993). *Helv. Chim. Acta.* **76**, 1435-1449.

Seela, F., Mersmann, K., Grasby, J. A. and Gait, M. J. (1993). *Helv. Chim. Acta* **76**, 1809-1820.

Shin, C., Choi, J. N., Song, J. T., Ahn, J. H., Lee, J. S. and Choi, Y. D. (1996). *Nucleic Acids Res.* **24**, 2685-2689.

Sigurdsson, S., Tuschl, T. and Eckstein, F. (1995). *RNA* **1**, 575-583.

Sigurdsson, S. T. and Eckstein, F. (1996). *Nucleic Acids Res.* **24**, 3129-3133.

Sinha, N. D., Davis, P., Usman, N., Pérez, J., Hodga, R., Kremsky, J. and Casale, R. (1993). *Biochimie* **75**, 13-23.

Slim, G. and Gait, M. J. (1991). *Nucleic Acids Res.* **19**, 1183-1188.

Slim, G. and Gait, M. J. (1992). *Biochem. Biophys. Res. Comm.* **183**, 605-609.

Sproat, B., Colonna, F., Mullah, B., Tsou, D., Andrus, A., Hampel, A. and Vinayak, R. (1995). *Nucleosides and Nucleotides* **14**, 255-273.

Sproat, B. S., Iribarren, A. M., Garcia, R. G. and Beijer, B. (1991). *Nucleic Acids Res.* **19**, 733-738.

Stump, W. T. and Hall, K. B. (1995). *RNA* **1**, 55-63.

Sumner-Smith, M., Roy, S., Barnett, R., Reid, L. S., Kuperman, R., Delling, U. and Sonnenberg, N. (1991). *J. Virol.* **65**, 5196-5201.

Talbot, S. J., Goodman, S., Bates, S. R. E., Fishwick, C. W. G. and Stockley, P. G. (1990). *Nucleic Acids Res.* **18**, 3521-3528.

Tan, R. and Frankel, A. D. (1992). *Biochemistry* **31**, 10288-10294.

Tao, J. and Frankel, A. D. (1992). *Proc. Natl. Acad. Sci. USA* **89**, 2723-2726.

Tsou, D., Hampel, A., Andrus, A. and Vinayak, R. (1995). *Nucleosides and Nucleotides* **14**, 1481-1492.

Tuschl, T., Ng, M. M. P., Pieken, W., Benseler, F. and Eckstein, F. (1993). *Biochemistry* **32**, 11658-11668.

Tuschl, T., Gohlke, C., Jovin, T. M., Westhof, E. and Eckstein, F. (1994). *Science* **266**, 785-789.

Usman, N. and Cedergren, R. (1992). *Trends Biochem. Sci.* **17**, 334-339.

Walter, N. G. and Burke, J. M. (1997). *RNA*, **3**, 392-404.

Wang, Z. and Rana, T. M. (1995). *J. Am. Chem. Soc.* **117**, 5438-5444.

Wang, Z. and Rana, T. M. (1996). *Biochemistry* **35**, 6491-6499.

Wang, Z., Wang, X. and Rana, T. (1996). *J. Biol. Chem.* **271**, 16995-16998.

Weeks, K. M., Ampe, C., Schultz, S. C., Steitz, T. A. and Crothers, D. M. (1990). *Science* **249**, 1281-1285.

Weeks, K. M. and Crothers, D. M. (1991). *Cell* **66**, 577-588.

Wincott, F., DiRenzo, A., Shaffer, C., Grimm, S., Tracz, D., Workman, C., Sweedler, D., Gonzalez, C., Scaringe, S. and Usman, N. (1995). *Nucleic Acids Res.* **23**, 2677-2684.

3

The Ribosome — the Universal Protein-synthesizing Machine[*]

RICHARD BRIMACOMBE

*Max-Planck-Institut für Molekulare Genetik, AG-Ribosomen,
Ihnestrasse 73, 14195-Berlin, Germany*

Abstract

In order to be able to understand the intricate processes involved in the biosynthesis of proteins on ribosomes at the molecular level, it is essential to obtain a correspondingly detailed knowledge of the structure of the ribosome itself. A significant advance in this direction has recently been provided by cryo-electron microscopy of *Escherichia coli* ribosomes combined with angular reconstruction techniques, which has led to computerized reconstructions of these ribosomes at a resolution of 20 Å. The reconstructions exhibit two features that are directly relevant to the three-dimensional folding *in situ* of the rRNA molecules. First, at this level of resolution, many fine structural details are visible, a number of them having dimensions comparable to those of nucleic acid helices. Second, in reconstructions of ribosomes in the pre- and post-translocational states, density can be seen which corresponds to the A and P site tRNAs and to the P and E site tRNAs, respectively. In the pre-translocational state the P site tRNA lies directly above the bridge connecting the two ribosomal subunits, with the A site tRNA fitted snugly against it in an 'S' configuration. The A site rRNA lies on the L7/12 side of the ribosome, and the angle between the planes of the two molecules is about 50°, in good agreement with many lines of biochemical evidence (but in disagreement with another recent electron microscopic study (Agrawal *et al*., 1996)). In the post-translocational state, the P site is again located above the intersubunit bridge, with the E site rRNA lying further towards the L1 side of the ribosome. These observations enable the decoding site on the 30 S ribosomal subunit to be located rather precisely. Accordingly, we have refined our previous model for the 3D folding of the *E. coli* 16 S contour. The relevant biochemical evidence

---

[*]All of the figures for this chapter appear in the colour plate section between p. 48 and p. 49.

THE MANY FACES OF RNA
ISBN 0-12-233210-5

includes new site-directed cross-linking data at the decoding centre, which define sets of contacts between the 16 S rRNA and mRNA or between 16 S rRNA and rRNA at the A, P or E sites; these data can be correlated directly with the electron microscopic model. Older intra-RNA cross-links within the 16 S rRNA itself are used to constrain other regions of the well-established 16 S secondary structure into three dimensions. The large body of available RNA–protein cross-linking and foot-printing data is also considered in the model, in order to correlate the rRNA folding with the known distribution of the 30 S ribosomal proteins as determined by neutron scattering and immuno-electron microscopy. The result is a new 3D model for the 16 S rRNA which fits the electron microscopic structure and at the same time satisfies the great majority of the known biochemical constraints.

1 Introduction

In every living cell the genetic information, in the form of mRNA, is translated into protein on the ribosome. The ribosome is itself composed of protein and RNA molecules, and is a massive particle which, even in the case of the relatively compact bacterial ribosome, has a total molecular weight of over two million daltons. Apart from its central role in the molecular biological function of the cell, the ribosome is also an important target for therapeutic intervention, because many antibiotics are known to interfere with the various steps of the protein biosynthetic cycle. In order to understand the intricate processes involved here, it is obviously essential to obtain a correspondingly detailed knowledge of the structure of the ribosome itself, and a great deal of effort has been expended on this topic over the last three decades, concentrating on the ribosome from the eubacterium *Escherichia coli*. Unfortunately, the huge size of the ribosome, combined with its structural heterogeneity at different stages of the functional cycle, make it an almost impossible target for classical X-ray crystallographic studies. Structural research has in consequence up to now been limited to low-resolution studies, largely aimed at determining the locations relative to one another of the various ribosomal components.

In the case of the ribosomal proteins, the most comprehensive study has been the derivation by neutron scattering techniques of a three-dimensional map defining the positions of the individual mass centres of the 21 proteins from the small (30 S) ribosomal subunit (Capel *et al.*, 1988). In the case of the ribosomal RNA (rRNA), which comprises two-thirds of the mass of the bacterial ribosome, secondary structure maps have been derived by sequence comparison methods for both the 16 S molecule from the small ribosomal subunit, and for the 23 S and 5 S molecules from the large (50 S) subunit (see e.g. Noller *et al.*, 1995; Brimacombe, 1995). These rRNA secondary structures can be correlated with the spatial distribution of the ribosomal proteins, with the help of RNA–protein interaction data obtained by cross-linking or foot-printing experiments (e.g. Osswald *et al.*, 1987; Powers and Noller, 1995). As a result, a number of models have been proposed (e.g. Brimacombe *et al*., 1988; Stern *et al*., 1988; Malhotra and Harvey, 1994) describing the three-dimensional arrangement of the helical elements of the rRNA molecules. Since these models have no quantitative physical basis apart from the distances in the neutron map for the protein mass centres just mentioned, they are essentially no more than 'cartoons' of the ribosomal structure. Very recently, however, high-resolution electron microscopic reconstructions of the ribosomal particles have become available, which have opened a new chapter in the study of the structure and function of the ribosome.

2 The new era in electron microscopy

2.1 Cryoelectron microscopy and image processing

Whereas older electron microscopic (EM) methodology for the study of biological specimens relied on negative staining combined with rather drastic sample preparation procedures, in the cryo-EM method the samples are simply flash frozen in solution. In the resulting matrix of vitreous ice the individual particles are randomly oriented, and this property is exploited in the computerized reconstruction of three-dimensional images of the particles (Frank *et al*., 1995; Serysheva *et al*., 1995). Thus, no tilting of the specimen is required, and several thousand images are considered, using pattern-recognition approaches to define classes of views. The 3D reconstruction of the particle derived from these views is subjected to iterative refinement procedures. So far no direct theoretical resolution limits for this single-particle approach are in sight (Henderson, 1995), and in the case of the bacterial ribosome the current resolution of the structures obtained has reached a level of c. 20 Å (Stark *et al*., 1997).

As already noted, two-thirds of the bacterial ribosome consists of rRNA, and about 60% of the rRNA is organized in double helices. Thus, these helices comprise 40% of the ribosomal mass, and, due to their high density, they would be expected to dominate the EM structures. The diameter of a nucleic acid helix is c. 20 Å, and, not surprisingly, in the latest EM reconstructions at 20 Å resolution, fine structural details are visible which can be correlated with individual helical elements of the rRNA.

2.2 Direct visualization of ribosomal ligands

A further expectation from the cryo-EM reconstructions at the current level of resolution is that it should be possible to see ligands such as tRNA molecules, as well as to be able to differentiate between ribosomes in various functional states. The ribosome has three binding sites for tRNA, the A or acceptor site for aminoacyl tRNA molecules arriving at the ribosome as specified by the codons on the mRNA, the P or peptidyl site for the tRNA carrying the already synthesized peptide chain, and the E or exit site for the empty tRNA that has passed on its peptide to the next tRNA molecule. The tRNAs, together with mRNA, move from one site to the next during the translocation step, with the result that in the pre-translocational state the ribosome carries two tRNA molecules at the A and P sites, respectively, and in the post-translocational state it carries tRNA molecules at the P and E sites. In EM reconstructions of ribosomes in these two states, it has indeed proved possible to visualize the tRNA molecules in the three ribosomal sites directly (Stark *et al.*, 1997; cf. Agrawal *et al.*, 1996). The A and P site tRNAs lie close together with an angle of c. 50° between the planes of the molecules, in agreement with biochemical predictions (Paulsen *et al.*, 1983; Smith and Yarus, 1989), whereas the E site tRNA lies somewhat further away (Stark *et al.*, 1997).

The location of the P site tRNA in the EM reconstruction of ribosomes in the post-translocational state is illustrated in Fig. 1. The figure shows the post-translocational 70 S ribosome as a semi-transparent gold-coloured structure, with the 30 S subunit from vacant ribosomes (carrying no tRNA) superimposed as a red non-transparent moiety. A 'knob' of density corresponding to the P site tRNA can be seen in the centre of the figure, to which the atomic structure of the tRNA molecule has been added. The lack of visible density in the EM reconstruction in the region of the tRNA acceptor stem (the lower right part of the tRNA molecule in Fig. 1) is most likely due to some flexibility in the tRNA binding (Stark *et al.*, 1997). Only the P site tRNA is visible in Fig. 1, the E site molecule being out of sight at the rear of the structure in this view. The positions of the A and P site tRNA molecules as visualized in the EM reconstructions are particularly important, because these positions precisely define the location of the decoding region of the 30 S ribosomal subunit, where the codons on the mRNA interact by Watson-Crick pairing with the corresponding anticodons on the tRNA; in Fig. 1 the P site tRNA anticodon is just out of sight on the left side of the tRNA molecule, where it penetrates the red contour of the 30 S subunit.

3   Three-dimensional modelling of the ribosomal RNA

3.1   Biochemical constraints on the rRNA folding

A number of biochemical data sets are available which give direct information concerning sites on the 16 S rRNA that are part of the decoding region of the 30 S subunit. These include cross-links to 16 S rRNA from specific positions on tRNA bound to the ribosomal A, P or E sites (summarized in Rinke-Appel *et al.*, 1995), corresponding foot-prints from tRNA at the A or P sites (Moazed and Noller, 1990), and cross-links from specific positions on mRNA close to the codons at the decoding site (Rinke-Appel *et al.*, 1994; Sergiev *et al.*, 1997). The three-dimensional locations of these cross-link and foot-print sites on the 16 S rRNA can now obviously be correlated with the corresponding locations of the tRNA and mRNA molecules on the 30 S subunit, as defined by the EM reconstructions just described. These tRNA and mRNA data thus serve as rigorous constraints on the folding of the 16 S rRNA molecule, in addition to the constraints already known from previous studies. The cross-linking and foot-printing data relating to the individual ribosomal proteins have already been mentioned above (Brimacombe et al., 1988; Stern et al., 1988), and other data sets relevant to the folding of the rRNA include sites of intra-RNA cross-linking within the 16 S rRNA (e.g. Stiege et al., 1986), cross-links at the subunit interface between the 16 S and 23 S rRNA molecules (Mitchell et al., 1992), and the immunoelectron microscopic (IEM) locations on the surface of the 30 S subunit of individual nucleotides within the 16 S rRNA (e.g. Stöffler and Stöffler-Meilicke, 1986). Not least, the phylogenetically established secondary structure of the rRNA itself, which contains several pseudoknot elements as well as numerous complex long-range interactions, severely constrains the three-dimensional folding of the molecule. Apart from these biochemical considerations, the fine structure of the EM reconstruction now provides a precise physical framework into which the rRNA molecules have to be fitted.

3.2 Fitting the rRNA model to the EM contour

Figure 2 shows the 16 S rRNA, modelled so as to fit into the 30 S moiety of the EM reconstruction of 70 S ribosomes in the pre-translocational state (Stark et al., 1997), carrying tRNAs at the A and P sites. The 16 S molecule is colour-coded white for double-helical regions, red for inter-helical single strands, green for intra-helical loops or bulges, and blue for hairpin loop ends. The model depicted in Fig. 2 represents the end product of a number of cycles of refinement, in which our older model (Brimacombe, 1995), derived before the EM reconstructions became available, has been progressively modified so as to fit the most recent EM contour, while at the same time satisfying as much of the biochemical data as possible. As in previous structures, the 30 S subunit is subdivided into three distinct domains, namely the 'head' (at the top of Fig. 2), the 'body' (centre and lower right) and the 'side lobe' or 'platform' (centre left). Particularly noteworthy in the EM reconstruction is the intricate pattern of bridges and tunnels in the central part of the structure at the junction of the three domains, which corresponds to the central connecting regions within the secondary structure of the 16 S molecule and also encompasses the decoding region (lying at the rear side of the subunit in this view). The head and side lobe regions of the subunit are noticeably less densely packed with rRNA than the body, an observation which is in agreement with the known distribution of protein and rRNA in the respective domains.

There is of course no space in this short article to give a detailed description of the rRNA model, and the extent to which it is compatible with the biochemical data; a full description has appeared in three publications (Mueller and Brimacombe, 1997a,b; Mueller et al., 1997). In the following sections, a few examples have been chosen so as to demonstrate the construction of the model in relation to the biochemical facts, and to illustrate some of the principal conclusions to which the model-building has led.

4 RNA–protein interactions

Up to now, X-ray crystallographic structures for three of the small ribosomal subunit proteins have been determined, namely for proteins S5, S8 and S17 (Ramakrishnan and White, 1992; Golden *et al*., 1993; Davies *et al*., 1996). In the absence of structural data for the remaining 18 proteins, the proteins in the neutron scattering map of Capel *et al*. (1988) are conventionally represented as spheres. As already mentioned above, the neutron map defines the positions of the mass centres of the 21 small subunit proteins. Furthermore, it is known, by comparison with IEM studies (Stöffler and Stöffler-Meilicke, 1986), which proteins within the map correspond to the major domains (head, body and side lobe) of the 30 S subunit. The neutron map, as an arrangement of protein spheres, can therefore be positioned within the EM reconstruction of the 30 S subunit without reference to the rRNA model. The approximate locations of the individual proteins and the corresponding RNA–protein cross-linking and foot-printing data already mentioned were of course taken into account during the building of the 16 S rRNA model, but nevertheless, in view of the large number of independent constraints on the rRNA folding, including both the fine structure of the EM reconstruction and the neutron map, it is important to check how successfully the RNA–protein data are accommodated in the final model. An example, for ribosomal protein S8, is shown in Fig. 3.

Figure 3 is a close-up view of the lower central region on the left side of Fig. 2, and the neutron position of S8 is represented by the red sphere. For better visibility of the rRNA environment, the diameter of the protein sphere has been reduced, so that it only represents c. one-third of the volume that would actually be occupied by a protein of this size. The rRNA elements that are included in Fig. 3 (see Brimacombe, 1995, for helix numbering) are those where RNA–protein interaction data for S8 have been documented, and the sites concerned are indicated by the coloured ball-and-stick nucleotides. The blue nucleotides in helix 21 are cross-link sites from the protein (Wower and Brimacombe, 1983), and it should be noted that the binding site of S8 has also been localized to the latter helix (Allmang *et al*., 1994). The yellow nucleotides are hydroxyl radical foot-print sites for S8 (Powers and Noller, 1995) and the red nucleotides are sites of base-specific foot-printing (Stern *et al*., 1988). It can be seen that, bearing in mind the reduced size of the protein sphere, the RNA–protein interaction data fit satisfactorily around the neutron position of the protein.

The corresponding fit to the RNA–protein interaction data for the majority of the small subunit proteins is similar (Mueller and Brimacombe, 1997b), although there are some exceptions. The most serious inconsistencies involve proteins, notably S13, S19 and S20, where there is some doubt as to the accuracy of their placement within the neutron map (see Brimacombe, 1995, for discussion), and in these cases the locations of the protein cross-link and foot-print sites in the rRNA model correspond well with the locations of these proteins as determined by IEM. Here it is important to note that, whereas the 16 S rRNA model is fitted to an EM reconstruction of 70 S ribosomes carrying tRNA, the neutron data were obtained with vacant 30 S subunits reconstituted from 16 S rRNA and isolated ribosomal proteins. This raises the question, crucial in model-building studies, as to how far discrepancies are due to errors (for whatever reason) within the individual data sets, and how far they are due to conformational differences between the respective ribosome preparations used to conduct the experiments. As the resolution of the rRNA models improves, this type of distinction will become progressively more significant.

5 The decoding region of the 30 S subunit

5.1 Foot-prints and cross-links to tRNA

The primary function of the 30 S subunit is the decoding of the mRNA, and so it is not surprising that the decoding region occupies a central position within the subunit. This is illustrated in Fig. 4, which shows the helical elements of the 16 S rRNA (represented as cylinders), together with the ribosomal proteins (as spheres), the A and P site tRNA molecules (light blue and green phosphate backbone tube models, respectively), and a segment of mRNA (the white backbone tube). The view in Fig. 4 is from the interface side of the subunit, that is to say rotated c. 180° about the vertical axis in relation to Fig. 2. The rRNA helices are coloured according to their locations within the 16 S secondary structure (cf. Brimacombe, 1995), dark blue for the 5'-domain, red for the central domain, light blue for the 3'-domain, and yellow for the 3'-terminal minor domain. The proteins (again reduced in size as in Fig. 3) interacting with the first three of these domains are correspondingly coloured darker blue, orange and blue-green, respectively, there being no proteins associated with the 3'-terminal minor domain. It can be seen that the decoding region, defined by the sites of interaction between the anticodon loops of the A and P site tRNA molecules and the mRNA, lies in the centre of the subunit, just where the four different rRNA domains come together.

Foot-print sites (Moazed and Noller, 1990) and cross-link sites (Rinke-Appel et al., 1995) for tRNA have in fact been observed in all four domains of the 16 S rRNA, and these data are accommodated satisfactorily in the rRNA model, with the notable exception of the foot-print and cross-link sites for P site tRNA in helix 23. The latter helix is located in the central domain of the 16 S rRNA, and can be seen in Fig. 4 as the red cylinder on the extreme right-hand side. While this helix is quite close to the position of the E site tRNA (not shown in Fig. 4), it is far away from the A and P site tRNA positions. The location of helix 23 in the rRNA model is determined by the fine structure of the EM reconstruction, and also by the locations of the ribosomal proteins in the side lobe region of the 30 S subunit (orange in Fig. 4). On the other hand the locations of the A and P site tRNA molecules are those directly visualized in the EM reconstruction of 70 S ribosomes in the pre-translocational state (Stark et al., 1997). This discrepancy is thus a case which can only be explained in terms of a different configuration of the ribosomes under the experimental conditions used to carry out the foot-printing and cross-linking experiments, and there is evidence from another EM reconstruction under appropriate conditions which strongly suggests that such a conformational change does indeed occur; this is discussed in detail in Mueller *et al.* (1997).

5.2   The path of the mRNA

An important consequence of the 16 S rRNA model is that it makes a clear prediction concerning the path followed by the mRNA through the 30 S subunit. Here, the experimental evidence consists of cross-links to the 16 S rRNA from thionucleotides (thiouridine or thioguanosine) introduced at specific sites within synthetic mRNA analogues, both upstream and downstream from the AUG initiator codon (Rinke-Appel *et al.*, 1994; Sergiev *et al.*, 1997). These cross-links are 'zero-length' cross-links, formed directly from the thionucleotide concerned, and they thus give particularly precise information on the neighbourhood of the mRNA relative to 16 S rRNA. At upstream sites in the mRNA, in the spacer region separating the AUG initiator codon from the Shine-Dalgarno sequence (which is complementary to the extreme 3'-terminus of the 16 S rRNA and interacts with the latter during the initiation phase of protein synthesis), three different cross-links to 16 S rRNA were observed. The cross-link formation was only partially dependent on the presence of the initiator tRNA^{fmet} and the same three cross-links were seen in varying amounts from all mRNA positions between one and eight nucleotides 5' to the AUG codon. This lack of specificity suggests that the upstream region of the mRNA (which has already been translated) is somewhat flexible.

In contrast, the cross-links from the downstream region of the mRNA were highly specific with regard to the mRNA position, and all of the cross-links found are to universally conserved nucleotides in the 16 S rRNA (Sergiev *et al.*, 1997). Furthermore, in all cases the cross-link formation was entirely dependent on the presence of initiator tRNA^{fmet}. The individual cross-links were from positions +2, +4, +6, +7, +8, +9, +11 and +12 of the mRNA, position +1 being the A-residue of the AUG initiator codon. These data are illustrated in Fig. 5, in which the mRNA is represented as a striped backbone tube model, with the P site tRNA molecule included (as a green backbone tube model) for reference. The cross-linked nucleotides on the 16 S rRNA are denoted by ball-and-stick nucleotides in the colour corresponding to their cross-linked partner on the mRNA. It is not possible to display all of the sites in one single view of the structure, but nevertheless a number of the cross-links can be seen in Fig. 5. These are the cross-links from position +12 of the mRNA to nucleotide 530 of the 16 S rRNA (red), from +11 to 532 (orange), from +8 or +9 to 1196 (yellow), and from +4 to 1402 (light blue). The other sites (where only the cross-linked target on 16 S rRNA is visible in Fig. 5) are from mRNA positions +7 (green), +6 (dark

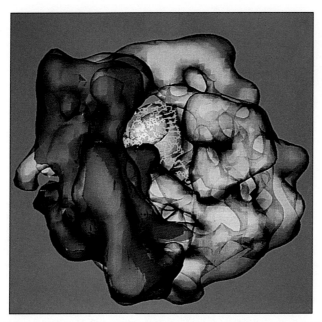

Chapter 3, Fig. 1 Location of the P site tRNA in the 70S ribosome. The 70S ribosome in the post-translocational state (Stark *et al.*, 1997) is shown as a semi-transparent envelope in gold, superimposed on a non-transparent silhouette in red of the 30S subunit from vacant ribosomes (Stark *et al.*, 1995). Areas of darker red occur where the 30S contour protrudes through the semi-transparent 70S contour. Due to the transparent rendering, the hue of the 50S subunit depends on how many 'layers' of the envelope are visible at any position; the structure takes on a progressively more golden appearance as the number of layers increases. The atomic structure of the P site tRNA is added in green-blue; here also the colour of the tRNA structure changes in those areas lying behind the semi-transparent 70S contour. The view of the ribosome is from the 'L7/L12' side.

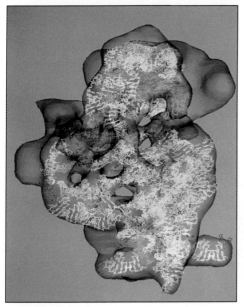

Chapter 3, Fig. 2 Fitting the rRNA to the EM reconstruction. The semi-transparent 30S subunit, viewed from the solvent side, has been cut out from the 70S EM reconstruction, and the 16S rRNA model is shown superimposed. White sequences in the 16S molecule are double helices, red sequences are inter-helical single strands, green sequences intra-helical loops and bulges, and blue sequences hairpin loop ends.

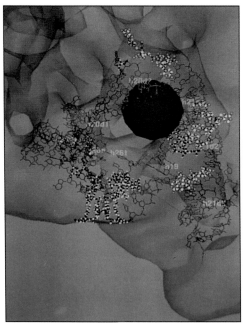

Chapter 3, Fig. 3 The interaction of the 16S rRNA with protein S8. The figure is a close-up of the lower centre area of the 30S subunit (Fig. 2), with S8 added as a red sphere. 16S rRNA helices are included which interact with the protein (see Brimacombe, 1995; Mueller and Brimacombe, 1997a, for helix numbering). Cross-link sites to the protein are indicated by blue ball-and-stick nucleotides, hydroxyl radical foot-print sites by yellow nucleotides, and base-specific foot-print sites by red nucleotides; see text for references.

Chapter 3, Fig. 4 The semi-transparent 30S subunit from the EM reconstruction, together with helical regions of the 16S rRNA (cylinders), ribosomal proteins (spheres), and the functional complex of A and P site tRNAs and mRNA (backbone tube models). The view is from the interface side of the subunit. Note that the protein spheres are of reduced size (cf. Fig. 3). See text for description of the colour coding. (The small black areas are software artifacts caused by coincidence in the computer rendering of the EM contour and of the spheres and cylinders.)

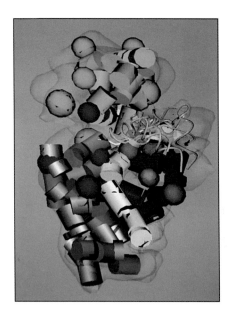

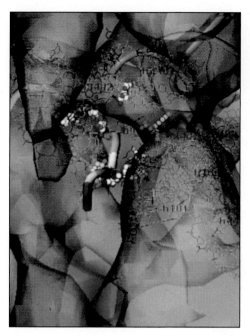

Chapter 3, Fig. 5 Path of the mRNA through the 30S subunit. The figure shows the head and upper body of the 30S subunit, rotated c 45° in relation to Fig. 1. The P site tRNA is included for reference (green backbone tube model), and the mRNA is a striped backbone tube. 16S rRNA helices are included which contain cross-links to the downstream region of the mRNA (Sergiev *et al.*, 1997), with the cross-link sites highlighted as ball-and-stick nucleotides in colours corresponding to the stripes on the mRNA molecule; see text for description of the colour coding, and for further details.

Chapter 3, Fig. 6 Locations of modified bases in the 16S rRNA. The semi-transparent 30S subunit is shown in a similar orientation to that of Fig. 1, with the A and P site tRNAs (light blue and green, respectively) and mRNA (white) added for reference. 16S rRNA regions containing modified nucleotides are added, with the actual positions of the modified nucleotides highlighted in CPK format. The colour coding is: pseudo-U-516 black, mG-527 red, mG-966 and mC-967 orange, mG-1207 yellow, mC-1402 green, mC-1407 light blue, mU-1498 dark blue, mG-1516 and m_2A-1518/1519 purple.

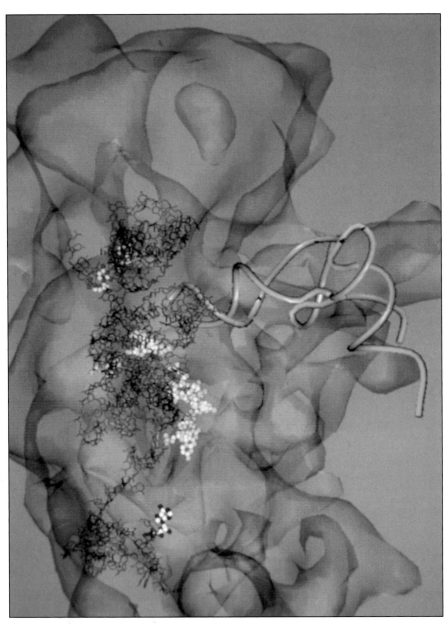

Chapter 3, Fig. 7 Locations of antibiotic interaction sites in the 16S rRNA. The orientation of the figure is similar to that of Fig. 6, with the A site tRNA included for reference. 16S rRNA helices are added which contain foot-print or resistance sites for the antibiotics streptomycin (yellow ball-and-stick nucleotides), spectinomycin (purple) and tetracycline (red); see text for references.

blue) and +2 (purple). The last of these cross-links was from a diazirine reagent attached to position +2 of the mRNA (Sergiev *et al*., 1997).

The orientation of the structure in Fig. 5 corresponds to a rotation of c. 45° about the vertical axis in an anti-clockwise direction viewed from above, in relation to the view shown in Fig. 1. It can be seen that the downstream (or incoming) mRNA passes through a hole in the 30 S subunit at the head–body junction of the latter (cf. Stark *et al*., 1997). Since the mRNA cross-linking data extend out to position +12, this is an excellent qualitative agreement with the 'toe-printing' data of Hartz *et al*. (1989), which indicate that the ribosome has a sharp 'boundary' corresponding to mRNA position +15. However, at the same time this mRNA arrangement immediately poses a problem. Namely, the initial contact between an mRNA molecule and the ribosome is the formation of the Shine-Dalgarno interaction with the 3′-terminus of the 16 S rRNA, which involves the upstream region of the mRNA relative to the AUG codon. (The upstream mRNA can be seen as the white non-striped part of the mRNA tube on the left in Fig. 5.) It thus seems inconceivable that, prior to forming this interaction, the mRNA should have to thread its way through a hole in the 30 S subunit. There is in fact only one covalent contact between the head and the body of the 30 S subunit (helix 28 of the 16 S rRNA (Brimacombe *et al*., 1988; Stern *et al*., 1988)), and the hole in the 30 S subunit can be more properly described as being the result of a second (non-covalent) contact between the two domains of the subunit; this second contact can be seen to the right of the mRNA molecule in Fig. 5, whereas the helix 28 covalent link lies further to the left (cf. Fig. 2). It is plausible to propose that the second contact is able to open or close in a manner analogous to the clamp or ring structures that have been observed in other systems such as DNA polymerase III (Herendeen and Kelly, 1996) or RNA polymerase (Polyakov *et al*., 1995). Thus, the mRNA could form the Shine-Dalgarno interaction in the 'open' phase, and the head–body junction would then close, clamping the mRNA into position. This would ensure that the mRNA is correctly located as it approaches the decoding area, and is consistent with the concept of a 'stand-by' site for the mRNA (Canonaco *et al*., 1989). Similar conclusions have been reached by other authors (Frank *et al*., 1995; Lata *et al*., 1996), albeit without the supportive evidence of the mRNA cross-link data and the three-dimensional model of the 16 S rRNA.

5.3 Clustering of modified nucleotides

The 16 S rRNA contains a number of post-transcriptionally modified nucleotides (Carbon *et al*., 1979; Van Charldorp *et al*., 1981; Bakin *et al*., 1994). These are distributed over nearly one thousand nucleotides in the primary sequence, but it has for some time been clear that their locations are closely related to the decoding areas of the 16 S molecule (Brimacombe *et al*., 1993). Not surprisingly, therefore, the modified bases form a compact cluster in the 3D model of the 16 S rRNA, and this is illustrated in Fig. 6. The orientation of the figure is approximately the same as that of Fig. 1, and the modified nucleotides (shown as coloured ball-and-stick nucleotides) all lie close to the subunit interface in the rRNA model, forming a 'cage' around the A and P site tRNA molecules.

It is well known that tRNA contains many similarly modified bases, and it seems likely that the modified bases in 16S rRNA will interact with those in the tRNA, so as to 'fine tune' the tRNA binding and decoding processes. We are, however, still a long way from even beginning to understand the molecular mechanisms involved here.

6   Sites of interaction with antibiotics

In general, two types of approach have been used to study the interaction of antibiotics with rRNA. The first is the foot-printing method (e.g. Moazed and Noller, 1987), which is analogous to that used with tRNA or ribosomal proteins, and the second is the analysis of mutations causing resistance to the antibiotic in question (e.g. Sigmund et al., 1984). As with the modified bases, the sites identified within the small ribosomal subunit by both these approaches correlate strongly with the decoding areas in the 16S rRNA. Indeed the foot-print sites for many antibiotics coincide precisely with the foot-print sites for either A or P site tRNA (Woodcock et al., 1991). In other cases the correlation is not quite so obvious, but nevertheless the 16S rRNA model indicates a close juxtaposition between the antibiotic sites and the decoding region of the 30S subunit. The data for the antibiotics streptomycin, spectinomycin and tetracycline are shown in Fig. 7, which also includes the A site tRNA as a backbone tube model for reference.

In the case of streptomycin, a number of resistance sites have been identified (summarized in Powers and Noller, 1991). These are spread over 900 nucleotides of the 16S rRNA, which in the secondary structure are concentrated into two regions, namely helices 1 and 18, and the pseudoknots (helices 2 and 18t, respectively) in each of the two areas. The foot-print sites for streptomycin are in the neighbourhood of the pseudoknot helix 2 (Moazed and Noller, 1987). Despite this spread, the 16S rRNA model shows all the streptomycin sites (depicted as yellow ball-and-stick nucleotides in Fig. 7) as a compact cluster just below the anticodon loop of the A site tRNA. In the case of spectinomycin, the respective foot-print and resistance sites (Sigmund et al., 1984; Moazed and Noller, 1987) are on opposite strands of the 16S helix 34, and these can be seen as the purple nucleotides above and to the left of the A site anticodon loop in Fig. 7. Sites of resistance to both spectinomycin and streptomycin have been identified in ribosomal protein S5 (Piepersberg et al., 1975), and in the crystal structure of the protein (Ramakrishnan and White, 1992) the respective sites lie on opposite sides of the molecule. In the 16S rRNA model the neutron map position of protein S5 is located just to the left of the cluster of streptomycin sites (in the orientation of Fig. 7), and the atomic structure of the protein can be oriented so as to bring the spectinomycin and streptomycin sites into the neighbourhood of their counterparts in the 16 S rRNA (unpublished data, but cf. Brimacombe, 1995).

The foot-print site for tetracycline (red in Fig. 7) lies in the single strand connecting helices 19 and 27 (Moazed and Noller, 1987), and it can be seen from Fig. 7 that this location is well below the A site tRNA. Helix 27 occupies this position in the model by virtue of an intra-RNA cross-link to helix 11, which is in the lower part of the 30 S subunit body (Mueller and Brimacombe, 1997a). However, there is a long single strand between helices 2 and 27 (visible in Fig. 7 between the lowest yellow streptomycin site and helices 19 and 27 at the bottom of the figure), and it is possible that in other conformational states the tetracycline foot-print site could be located closer to the decoding region.

7 Conclusions

The availability of the EM reconstructions at 20 Å resolution has for the first time provided a physical basis on which models for the three-dimensional arrangement of the rRNA can be built. Clearly, at the current level of resolution the 16 S rRNA model has not been finalized, and many improvements and modifications will surely need to be made as more biochemical data are accumulated and the resolution of the EM structures improves. Nevertheless, the fitting of the 16 S rRNA to the EM structure at the current level of resolution has already given us a new insight into the organization of this complex molecule and its interactions with functional ligands or antibiotics. As has been pointed out by Moore (1995), the application of EM reconstruction techniques is likely to be more fruitful than the X-ray crystallographic approach for some time to come. A particular advantage of the EM method is that it is well suited to the investigation of ribosomes in specific functional configurations as has already been exploited for the visualization of tRNA molecules on ribosomes in the pre- and post-translocational states (Stark *et al.*, 1997). Thus, conformational changes can be detected at different stages of the ribosomal cycle, and concepts developed, such as that described here for the path followed by the mRNA through the 30 S subunit. So far we have concentrated our attention on the modelling of the 16 S rRNA within the 30 S subunit. However, the 50 S subunit shows a very stable and detailed fine structure in the EM reconstructions, and a considerable amount of biochemical data is available (see Brimacombe, 1995, for review). The derivation of a new model for the 23 S rRNA, and hence for the complete rRNA content of the 70 S ribosome, is the next challenge.

References

Agrawal, R. K., Penczek, P., Grassucci, R. A., Li, Y., Leith, A., Nierhaus, K. H. and Frank, J. (1996). *Science* **271**, 1000–1002.
Allmang, C., Mougel, M., Westhof, E., Ehresmann, B. and Ehresmann, C. (1994). *Nucleic Acids Res.* **22**, 3708–3714.
Bakin, A., Kowalak, J. A., McCloskey, J. A. and Ofengand, J. (1994). *Nucleic Acids Res.* **22**, 3681–3684.

Brimacombe, R. (1995). *Eur. J. Biochem.* **230**, 365-383.

Brimacombe, R., Atmadja, J., Stiege, W. and Schüler, D. (1988). *J. Mol. Biol.* **199**, 115-136.

Brimacombe, R., Mitchell, P., Osswald, M., Stade, K. and Bochkariov, D. (1993). *FASEB J.* **7**, 161-167.

Canonaco, M A., Gualerzi, C. O. and Pon, C. L. (1989). *Eur. J. Biochem.* **182**, 501-506.

Capel, M. S., Kjeldgaard, M., Engelman, D. M. and Moore, P. B. (1988). *J. Mol. Biol.* **200**, 65-87.

Carbon, P., Ehresmann, C., Ehresmann, B. and Ebel, J. P. (1979). *Eur. J. Biochem.* **100**, 399-410.

Davies, C., Ramakrishnan, V. and White, S. W. (1996). *Structure* **4**, 1093-1104.

Frank, J., Zhu, J., Penczek, P., Li, Y., Srivastava, S., Verschoor, A., Radermacher, M., Grassucci, R., Lata, R. K. and Agrawal, R. K. (1995). *Nature* **376**, 441-444.

Golden, B. L., Hoffmann, D., Ramakrishnan, V. and White, S. W. (1993). *Biochemistry* **32**, 12812-12820.

Hartz, D., McPheeters, D. S. and Gold, L. (1989). *Genes Develop.* **3**, 1899-1912.

Henderson, R. (1995). *Quart. Rev. Biophys.* **28**, 171-193.

Herendeen, D. R. and Kelly, T. J. (1996). *Cell* **84**, 5-8.

Lata, K. R., Agrawal, R. K., Penczek, P., Grassucci, R., Zhu, J. and Frank, J. (1996). *J. Mol. Biol.* **262**, 43-52.

Malhotra, F. and Harvey, S. C. (1994). *J. Mol. Biol.* **240**, 308-340.

Mitchell, P., Osswald, M. and Brimacombe, R. (1992). *Biochemistry* **31**, 3004-3011.

Moazed, D. and Noller, H. F. (1987). *Nature* **327**, 389-394.

Moazed, D. and Noller, H. F. (1990). *J. Mol. Biol.* **211**, 135-145.

Moore, P. B. (1995). *Structure* **3**, 851-852.

Mueller, F. and Brimacombe, R. (1997a). *J. Mol. Biol.* **271**, 524-544.

Mueller, F., and Brimacombe, R. (1997b). *J. Mol. Biol.* **271**, 545-565.

Mueller, F., Stark, H., van Heel, M., Rinke-Appel, J. and Brimacombe, R. (1997). *J. Mol. Biol.* **271**, 566-587.

Noller, H. F., Green, R., Heilek, G., Hoffarth, V., Hüttenhofer, A., Joseph, S., Lee, I., Lieberman, K., Mankin, A., Merryman, C., Powers, T., Puglisi, E. V., Samaha, R. R. and Weiser, B. (1995). *Biochem. Cell Biol.* **73**, 997-1009.

Osswald, M., Greuer, B., Brimacombe, R., Stöffler, G., Bäumert, H. and Fasold, H. (1987). *Nucleic Acids Res.* **15**, 3221-3240.

Paulsen, H., Robertson, H. J. and Wintermeyer, W. (1983). *J. Mol. Biol.* **167**, 411-426.

Piepersberg, W., Böck, A., Yaguchi, M. and Wittmann, H. G. (1975). *Mol. Gen. Genet.* **143**, 43-52.

Polyakov, A., Severinova, E. and Darst, S. A. (1995). *Cell* **83**, 365-373.

Powers, T. and Noller, H. F. (1991). *EMBO J.* **10**, 2203-2214.

Powers, T. and Noller, F. (1995). *RNA* **1**, 194-209.

Ramakrishnan, V. and White, S. W. (1992). *Nature* **358**, 768-771.

Rinke-Appel, J., Jünke, N., Brimacombe, R., Lavrik, I., Dokudovskaya, S., Dontsova, O. and Bogdanov, A. (1994). *Nucleic Acids Res.* **22**, 3018-3025.

Rinke-Appel, M., Jünke, N., Osswald, M. and Brimacombe, R. (1995). *RNA* **1**, 1018-1028.

Sergiev, P., Lavrik, I. Wlasoff, V., Dokudovskaya, S., Dontsova, O., Bogdanov, A. and Brimacombe, R. (1997). *RNA* **3**, 464-475.

Serysheva, I. I., Orlova, E. V., Chiu, W., Sherman, M. B., Hamilton, S. L. and van Heel, M. (1995). *Nature Struct. Biol.* **2**, 18-24.

Sigmund, C. D., Ettayebi, M. and Morgan, E. A. (1984). *Nucleic Acids Res.* **12**, 4653-4663.

Smith, D. and Yarus, M. (1989). *Proc. Natl. Acad. Sci. USA* **86**, 4397-4401.

Stark, H., Mueller, F., Orlova, E. V., Schatz, M., Dube, P., Erdemir, T., Zemlin, F., Brimacombe, R. and van Heel, M. (1995). *Structure* **3**, 815–821.

Stark, H., Orlova, E. V., Rinke-Appel, J., Jünke, N., Mueller, F., Rodnina, M., Wintermeyer, W., Brimacombe, R. and van Heel, M. (1997). *Cell* **88**, 19–28.

Stern, S., Weiser, B. and Noller, H. F. (1988). *J. Mol. Biol.* **204**, 447–481.

Stiege, W., Atmadja, J., Zobawa, M. and Brimacombe, R. (1986). *J. Mol. Biol.* **191**, 135–138.

Stöffler, G. and Stöffler-Meilicke, M. (1986). In *Structure, Function and Genetics of Ribosomes* (B. Hardesty and G. Kramer, eds) pp. 28–46, Springer-Verlag, New York.

Van Charldorp, R., Heus, H. A. and van Knippenberg P. H. (1981). *Nucleic Acids Res.* **9**, 2717–2725.

Woodcock, J., Moazed, D., Cannon, M., Davies, J. and Noller, H. F. (1991). *EMBO J.* **10**, 3099–3103.

Wower, I. and Brimacombe, R. (1983). *Nucleic Acids Res.* **11**, 1419–1437.

4

tRNA Recognition by Aminoacyl-tRNA Synthetases*

STEPHEN CUSACK, ANYA YAREMCHUK AND MICHAEL TUKALO

European Molecular Biology Laboratory, Grenoble Outstation, c/o ILL, 156X, F-38042 Grenoble Cedex 9, France

Abstract

Crystal structures are now known of six aminoacyl-tRNA synthetase complexes and with their cognate tRNA. These structures are reviewed with regard to general features of protein–RNA recognition as well as to their different specific modes of tRNA recognition. Emphasis is given to results on seryl-tRNA synthetase, a class IIa synthetase which specifically binds to the long variable arm of tRNAser but does not interact with anticodon, to prolyl-tRNA synthetase, a class IIa synthetase (together with ThrRS, GlyRS and HisRS) with a C-terminal anticodon binding domain with a unique α/β-fold and lysyl-tRNA synthetase, a class IIb synthetase (together with AspRS and AsnRS) with an N-terminal anticodon binding domain containing an OB-fold. These examples suggest that the type of additional RNA binding domain coupled with the class II synthetase catalytic domain is correlated with the number of anti-codon nucleotides that have to be recognized to discriminate tRNAs.

1 Introduction

The fidelity of protein synthesis depends to a large extent on the high specificity with which aminoacyl-tRNA synthetases charge their cognate tRNAs with the correct amino acid. Synthetases catalyse the aminoacylation reaction in two steps, firstly the activation of the amino acid using ATP to form the enzyme bound aminoacyl-adenylate and secondly, the transfer of the amino acid to the 2' or 3' hydroxyl of the ribose of the 3' terminal A-76 of the tRNA. In *Esherichia coli*,

*All images for this chapter appear in the colour plate section between p. 112 and p. 113.

THE MANY FACES OF RNA
ISBN 0-12-233210-5

there are at least 46 different tRNA molecules with anticodons corresponding to the various amino acids, and the seryl-tRNA synthetase, for instance, has to selectively charge the six serine isoacceptors and ignore the others. Given that tRNAs superficially have similar secondary and tertiary structures, what is the structural basis for the specific recognition between aminoacyl-tRNA synthetases and tRNAs?

2   tRNA identity

In order to analyse this question, the concept of tRNA identity elements has been introduced (reviewed in Saks *et al*., 1994). This refers to those structural elements, usually a limited number of nucleotides or base pairs, that are in general conserved amongst tRNA iso-acceptors for an amino acid X and are essential for specific recognition by the cognate synthetase. In the ideal case, when these identity elements are transplanted into a tRNA for a different amino acid Y, they switch the identity of this tRNA so that it becomes an X iso-acceptor. However, negative identity elements also play an important role in tRNA discrimination and an identity switch experiment will not work ideally if there are negative elements for synthetase X remaining in tRNA Y. It is to be expected that the synthetase makes direct contacts to the tRNA identity elements and in fact crystal structures show that this is generally true. However, some identity elements function more subtly by locally modifying the tRNA structure which can indirectly affect contacts with the synthetase. Of course, as discussed below, synthetase contacts with their cognate tRNAs are more extensive than just with identity elements; additional contacts are important to increase binding affinity and to reinforce induced fit conformational changes.

In many amino acid systems, extensive biochemical studies have revealed the major tRNA identity elements, although for a particular system they are not necessarily conserved from one organism to another. In *E. coli*, 17 of the 20 aminoacyl-tRNA synthetases use some or all of the three anti-codon bases for specific recognition, the exceptions being seryl-, alanyl- and leucyl-tRNA synthetases (Saks *et al*., 1994). This makes sense, since the anticodon specifies the amino acid and indeed, in many cases, due to the degeneracy of the genetic code, the second two bases suffice. The absence of anticodon recognition in the serine system can be explained by the variety of serine anticodons, from two distinct groups, so that serine isoacceptors share no common anticodon base. The second most important region of tRNA where identity elements are found comprises the discriminator base 73 and the first few base pairs of the acceptor stem. These regions are close to the enzyme active site which is a crucial place to provide specific interactions which permit only the cognate tRNA to be correctly positioned for the amino acid transfer step. In addition to the anticodon and acceptor stem regions, idiosyncratic features of certain tRNAs are used as identity elements, such as the long variable arm of tRNA^{ser} (see below), the extra G-1:C73 base-pair in RNA^{his} (Himeno *et al*., 1989) and the unique G3:U70 base pair in tRNA[ala] (Hou and Schimmel, 1988). Aminoacyl-tRNA synthetases not only have to specifically recognize and bind cognate tRNAs but also to perform a two-step enzymatic reaction resulting in the esterification of either the $2'$ or $3'$ hydroxyl of the ribose of the terminal A-76 with cognate amino acid, followed by product release. This permits complex mechanisms of discrimination against charging non-cognate tRNAs, which need not be at the initial binding step, but can be at the catalytic step or indeed through hydrolysis of a mischarged tRNA by a specific editing activity of the synthetase (Lin *et al*., 1996).

3 The two classes of aminoacyl-tRNA synthetases

A more detailed picture of specific tRNA recognition and catalysis by aminoacyl-tRNA synthetases is emerging from crystallographic studies of aminoacyl-tRNA synthetases complexed with various combinations of their three substrates ATP, cognate amino acid and cognate tRNA and the activated amino acid intermediate, the aminoacyl-adenylate. In 1995, crystal structures were known of 11 of the 20 aminoacyl-tRNA synthetases (Cusack, 1995) and since then three new synthetase structures have been solved, asparaginyl-tRNA synthetase (Berthet-Colominas *et al*., unpublished), isoleucyl-tRNA synthetase (Nureki *et al*., unpublished) and prolyl-tRNA synthetase (Cusack *et al*., unpublished). It is interesting to note that ten different *T. thermophilus* synthetase structures have been determined, the increased thermal stability of these enzymes perhaps explaining the high success rate for their crystallization.

Evolution has resulted in two completely distinct structural solutions to the aminoacylation problem since it is found that the aminoacyl-tRNA synthetases are partitioned into two exclusive classes of ten enzymes each, class I and class II (Cusack *et al*., 1990; Eriani *et al*., 1990). The catalytic domain of class I enzymes contains the well-known Rossmann (nucleotide binding) fold as a framework whereas that of class II enzymes is based around a novel anti-parallel fold first revealed in the structure of seryl-tRNA synthetase (Cusack *et al*., 1990). The catalytic domain of each class includes short sequence motifs, 'HIGH' and 'KMSKS' in class I and motifs 1, 2 and 3 in class II and the extremely few absolutely conserved residues in each class are involved with ATP binding and catalysis of amino acid activation (Cavarelli *et al*., 1994; Belrhali *et al*., 1995). A functional distinction between the two classes is that class I enzymes charge the $2'$ hydroxyl and class II enzymes (except phenylalanyl-tRNA synthetase) charge the $3'$ hydroxyl of the ribose of A-76 (Fraser and Rich, 1975; Sprinzl and Cramer, 1975; Eriani *et al*., 1990). Class II synthetases are almost invariably functional dimers whereas most class I enzymes are monomers except for TyrRS and TrpRS.

4 RNA-binding modules in aminoacyl-tRNA synthetases

In both classes, the tRNA binding ability of the synthetase is augmented by RNA-binding modules which, because of their greater structural variability, have presumably been added to the catalytic domain at a later stage in evolution (Cusack *et al*., 1991; Delarue and Moras, 1993). These extra domains contribute to the elongated shape of the enzyme subunits thus permitting binding to distal regions of the asymmetric tRNA molecule where some 60 Å separates the anti-codon from the 3' end. Analysis of synthetase sequences and structures shows that sub-classes of synthetases share homologous anticodon binding modules. In addi-tion, there are often additional domains inserted into the catalytic domain which show even greater idiosyncrasy, and which are also involved to some extent in RNA binding. The modular structure, involving the class specific catalytic domain with conserved motifs, sub-class specific anticodon binding domains and system-specific insertion domains is most obvious in class II synthetases as illustrated by their linear domain structure (Fig. 1) and in three-dimensional structures (Figs 2 and 3). For instance four of the five class IIa synthetases (ThrRS, ProRS, GlyRS and HisRS, but not SerRS) possess a C-terminal anticodon binding module of about 100 residues which has a novel α/β fold so far not found in another protein (Fig. 2). By contrast, the three closely related class IIb synthetases (AspRS, AsnRS and LysRS, Fig. 3) all possess an N-terminal anticodon binding module, again of about 100 residues, but with a more widespread five-stranded β-barrel fold (oligomer binding or OB fold, Murzin, 1993). The OB-fold is found in a variety of other nucleic acid binding proteins, for example the cold-shock protein, several single-stranded DNA binding proteins (Bochkarev *et al*., 1997) and ribo-somal protein S1 (Bycroft *et al*., 1997). SerRS is unusual in having no anti-codon binding domain, but instead a remarkable long helical arm which contacts the tRNA as described in more detail below. Phenylalanyl-tRNA synthetase is a complex class II synthetase with an $(\alpha\beta)_2$ subunit architecture. The α-subunit contains the catalytic domain and the large β-subunit contains, among several other domains, both an OB-fold domain and an RNP domain (Mosyak *et al*., 1995). It turns out that the OB-fold domain is not involved in tRNA binding and it is the RNP domain which performs anticodon recognition (Goldgur *et al*., 1997). Furthermore a coiled coil, helical arm at the N-terminal of the α-subunit, similar to that which occurs in SerRS and which is disordered in the uncomplexed PheRS, is also important in tRNA binding (Goldgur *et al*., 1997).

In class I, a particularly interesting case is that of the glutaminyl- and glutamyl-tRNA synthetases which, apart from the anticodon recognition domains, are strik-ingly similar (Nureki *et al*., 1995). In GlnRS the anticodon is recognized by two β-barrel domains, whereas in GluRS, the spatially equivalent putative anticodon binding domains are entirely α-helical. This is one example of the less common, all α-helical RNA-binding modules, another being the anticodon binding domain found in the sub-class of class I synthetases containing methionyl- (Brunie *et al*., 1990), valyl- and isoleucyl-tRNA synthetases. Tyrosyl- and tryptophanyl-tRNA synthetases, two very closely related class I synthetases, also contain helical puta-tive anticodon binding domains (see e.g. Doublié *et al*., 1995). Although no crystal structure is available of a tRNA complex with either of these two dimeric class I enzymes, there is strong evidence that the tRNA must bind across both subunits.

5   General features of tRNA recognition by aminoacyl-tRNA synthetases

The three systems for which the most extensive structural data exists on protein-tRNA recognition are the class I glutaminyl-system (Rould *et al.*, 1989; Rould *et al.*, 1991) and the class II aspartyl- (Ruff *et al.*, 1991; Cavarelli *et al.*, 1993) and seryl-systems (Biou *et al.*, 1994; Cusack *et al.*, 1996a). These show strikingly different modes of specific synthetase-tRNA interaction but share certain general features. These include the presence of a large synthetase-tRNA interaction interface characterized by, firstly, non-specific backbone contacts, often involving basic residues, which both increase binding affinity as well as aiding correct positioning and orientation of the tRNA and secondly, discriminatory base-specific interactions restricted to a few regions at or close to identity elements, principally the anticodon and the acceptor stem. The second general feature is mutual induced fit by which protein–RNA contacts are made as a result of conformational changes in either or both macromolecules, emphasizing the fact that tRNA-synthetase recognition is a dynamic phenomenon. This includes ordering of protein loops and re-orientation and stabilization of domains (e.g. SerRS), base-pair breaking in the acceptor stem and 3′ end distortion (e.g. tRNA^{gln}), de-stacking of bases in the anticodon loop (e.g. tRNA^{gln}, tRNA^{asp}) or formation of additional, not necessarily Watson-Crick base-pairs (e.g. tRNA^{gln}). More recent crystallographic results have been obtained on three other class II synthetase-tRNA complexes from the lysyl- (Cusack *et al.*, 1996b), phenylalanyl- (Goldgur *et al.*, 1997) and prolyl-systems (Cusack *et al.*, unpublished) which confirm these general features. A previous review has focused mainly on the GlnRS and AspRS complexes (Arnez and Moras, 1994).

6   Glutaminyl-tRNA synthetase (GlnRS)

GlnRS is a monomeric class I synthetase whose specificity for tRNA^{gln} is largely determined by interactions with identity elements in the tRNA acceptor stem and anti-codon stem-loop, both of which have severe distortions from the structure presumed to exist in the uncomplexed tRNA (Rould *et al.*, 1989; Rould *et al.*, 1991). In the complex, the tRNA anticodon stem is extended from five to seven base pairs by two extra non Watson-Crick base pairs. The three anticodon bases (34-CUG) are splayed out to fit into three separate recognition pockets formed at the interface between the distal two β-barrel domains of the protein. In the active site of the synthetase, the tRNA is orientated in such a way that specific interactions can be made within the acceptor stem minor groove to identity determinants in the second and third base-pairs. On the other hand, the tRNA single-stranded 3′ end can only reach the catalytic centre by forming an unusual hairpin turn. This conformation requires the breaking of the first U1-A72 base-pair and is stabilized, in part, by a hydrogen bond between the discriminator base G-73 and the phosphate of A-72. The three class I systems GlnRS, GluRS and ArgRS are unique amongst synthetases in that bound cognate tRNA is required for the amino acid activation reaction although the structural basis for this dependence is not yet fully understood.

7 tRNA recognition by class IIb synthetases

Crystal structures are now known of the three class IIb synthetases (Fig. 3), AspRS (Ruff *et al.*, 1991; Delarue *et al.*, 1994), LysRS (Onesti *et al.*, 1995; Cusack *et al.*, 1996b) and AsnRS (Berthet-Colominas *et al.*, unpublished). The crystal structure of the complex of yeast AspRS with its cognate tRNA shows that the enzyme can symmetrically bind two tRNAs although each tRNA interacts predominantly with only one subunit (Ruff *et al.*, 1991). In contrast to the class I case, class II synthetases interact with the acceptor stem of their cognate tRNAs from the major groove side. Normally the major groove of regular A-form RNA is too deep and narrow to permit easy access by protein side-chains to bases, but at distortions in the helix due to mismatches or internal bulges as well as at helical extremities, the major groove is accessible. In class II synthetases, interaction within the major groove of the acceptor stem is achieved by means of the so-called motif 2 loop (see Cusack *et al.*, 1996b). In the case of yeast AspRS this makes base-specific interactions with the discriminator base G-73, a major identity element, as well as base-pair U1–A72. The major groove side recognition also means that the single-stranded 3′ end of the tRNA can enter the synthetase active site without significant distortion from its normal helical path, again a radical difference from the class I case.

Anticodon recognition by class IIb synthetases is performed by an N-terminal five-stranded β-barrel domain with an OB-fold. As shown by comparing the crystal structures of free and complexed yeast tRNA[asp], the normal compact structure of the free tRNA anti-codon loop, with stacked anticodon bases, undergoes a large conformational change upon binding to AspRS and the five anticodon loop bases are splayed out to the exterior. The three anticodon bases (34-GUC in the case of yeast tRNA[asp]) lie across the β-sheet surface and are recognized by specific hydrogen bonding interactions with the enzyme (Cavarelli *et al.*, 1993). Recent crystallographic results on another class IIb synthetase complexed with its cognate tRNA, *Thermus thermophilus* lysyl-tRNA synthetase (Cusack *et al.*, 1996b), show a very similar mode of interaction between the anti-codon of tRNA[lys](CUU) and the N-terminal β-barrel domain of LysRS (Fig. 4). Details of the interactions of LysRS with the three anti-codon bases of tRNA[lys] (CUU) are shown in Fig. 5. All tRNAs cognate to the three class IIb synthetases (AsnRS, AspRS and LysRS) contain a central U-35 which is in each case a major identity element. In LysRS and AspRS, U-35 is found to stack with a phenylalanine and make specific hydrogen bonding interactions with a glutamine and arginine (Fig. 5), all three of which are residues absolutely conserved in all class IIb synthetases. Base 36 is also an important identity element; in the case of AspRS, C-36 is recognized by main-chain interactions, and in LysRS, U-36 interacts specifically with conserved side-chains (Fig. 5). The wobble position 34 is important for the amino acid specificity of tRNAs cognate to class IIb synthetases, since they occupy only half codon groups in the genetic code. LysRS, for instance, has to recognize a C or modified U (e.g. commonly mnm^5s^2U) in position 34 (Fig. 5) and discriminate against a G in this position, GUU being the most frequent asparagine anticodon (the G often being hyper-modified to Q). The LysRS complex structure shows clearly that a Q-base, with its large modification at the N7 position, could not be accommodated without severe distortions, thus providing strong discrimination against $tRNA^{asn}$ (QUU) (Cusack *et al.*, 1996b). On the other hand, the glutamic acid residue in AspRS interacting specifically with the base G/Q-34 in $tRNA^{asp}$ (Cavarelli *et al.*, 1993) is conserved in AsnRS sequences suggesting a similar mode of recognition of this base in these two very similar class IIb synthetases.

Class IIb synthetases have a small but distinct 'hinge' domain between the N-terminal anti-codon binding domain and the C-terminal catalytic domain. Although the angular orientation of the anticodon binding domain relative to the catalytic domain is quite variable among class IIb synthetases (even for synthetases for the same amino acid but from different organisms; see discussion in Cusack *et al.*, 1996b), there is no evidence to show that there is significant relative movement of the two domains upon tRNA binding. In the yeast aspartyl-system, the hinge module makes contacts with the $tRNA^{asp}$ in the D-stem region, although not directly with the G10-U25 identity elements (Cavarelli *et al.*, 1993). In LysRS the interactions of the hinge region are not known but an extra N-terminal helix at the beginning of the anticodon binding domain makes contacts with the tRNA backbone in the region of nucleotides 26–27 (Cusack *et al.*, 1996b, Fig. 4).

8 tRNA recognition by class IIa aminoacyl-tRNA synthetases

Crystal structures of four class IIa synthetases are now known: SerRS (see below), HisRS (Arnez *et al.*, 1995; Åberg *et al.*, 1997), GlyRS (Logan *et al.*, 1995) and ProRS (Cusack *et al.*, unpublished), the last three of which all contain a homologous C-terminal anticodon binding domain (Fig. 2). Among these three enzymes, the position of the anticodon binding domain relative to the catalytic domain varies significantly, although in each case it roughly occupies the same spatial position as the N-terminal anticodon binding domain in class IIb synthetases. In HisRS the anticodon binding domain is connected to the catalytic domain by an extended peptide (Fig. 2) and exclusively packs against the catalytic domain of the *other* subunit in the dimer. In ProRS the anticodon binding domain exclusively packs against the catalytic domain of the *same* subunit and in GlyRS it packs against both subunits.

How does the tRNA interact with class IIa synthetases and in particular how does the C-terminal RNA-binding domain interact with the anticodon? In the case of *E. coli* (Arnez *et al.*, 1995) and *T. thermophilus* (Åberg *et al.*, 1997)

HisRS and *T. thermophilus* GlyRS (Logan *et al.*, 1995), tRNA docking models have been proposed based on electrostatic potential surfaces and knowledge of the mode of binding of tRNA to the class II synthetases, AspRS and SerRS. These all suggest that one tRNA will bind predominantly to one subunit but with some cross-subunit contacts in the D-stem region. In the case of *T. thermophilus* HisRS it is proposed that the helical domain inserted into the catalytic domain between motif 2 and motif 3 would clamp down onto the 3' strand of the acceptor stem of the tRNA (Åberg *et al.*, 1997). In *T. thermophilus* GlyRS an extra domain inserted into the catalytic domain between, unusually, motif 1 and motif 2, could interact with the tRNA acceptor stem from the minor groove side (Logan *et al.*, 1995).

Recent results on *T. thermophilus* ProRS give the first crystallographic data on the actual mode of binding of tRNA to a class IIa synthetase. Sequence analysis shows that there are two distinct structural forms of ProRS which must have diverged early in evolution: (a) 'eukaryote/archae-like' characterized by the absence of an insertion domain between motifs 2 and 3 and an extra C-terminal domain beyond the normal class IIa anticodon binding domain; and (b) 'prokaryote-like', which are larger enzymes with a very large insertion between motifs 2 and 3 and no extra C-terminal domain (Fig. 1). Surprisingly, *T. thermophilus* ProRS is of the 'eukaryote/archae-like' form. The crystal structure of the enzyme at 2.7 Å shows two novel features among class II synthetases (Fig. 6). The unique extra C-terminal domain is in fact a zinc-binding domain situated in the position normally occupied by the insertion domain between motifs 2 and 3. The zinc is tetrahedrally coordinated by four cysteine residues and appears to have a structural role. The Zn-domain is connected to the preceding class IIa anticodon binding domain by a long α-helix. The extreme C-terminus, an absolutely conserved tyrosine in the 'eukaryote/archae-like' ProRSs, doubles back into the active site strongly suggesting that the exposed carboxyl-group plays an important role in activity.

The crystal structure of the complex of *T. thermophilus* ProRS with tRNA^{pro} (CGG) has been solved at 3.5 Å resolution. Despite the modest resolution, the mode of binding of the anticodon stem-loop to the class IIa anticodon binding domain is clear, the main interacting surface being the β-sheet and an α-helix which approaches the anticodon loop from the major groove side (Fig. 7). The distortion introduced into the anticodon loop is reminiscent of that occurring upon cognate tRNA binding to class IIb synthetases, with bases G-35, G-36 and G-37 splayed out, but differs in that base C-34 remains stacked under base U-33 and bases U-32 and A-38 form a base pair (Fig. 7). In common with the class IIa synthetases GlyRS and ThrRS, ProRS has to recognize cognate tRNAs in which the second and third anticodon nucleotides uniquely define the amino acid, whereas the first (wobble) base is variable (i.e. NGG for tRNA^{pro} isoacceptors, NCC for tRNA^{gly} and NGU for tRNA^{thr}). The structure explains how this functional requirement is met by clearly showing specific recognition of bases G-35 and G-36 by side-chains from the synthetase, whereas base C-34 makes no interactions and is stacked under base U-33 in a pocket which could accommodate any nucleotide. More precise details of the interaction of tRNA^{pro} with ProRS await a higher resolution structure of the complex.

9 Seryl-tRNA synthetase (SerRS)

On the basis of primary sequence homology, SerRS is also classified as a class IIa synthetase (Cusack *et al.*, 1991) but it lacks the normal class IIa anticodon binding domain and its mode of tRNA recognition is unique. The variety of serine anticodons, from two distinct groups, means that the serine isoacceptors share no common anticodon base. That the anticodon is not recognized by SerRS, whereas the tRNA^{ser} long variable arm, a feature only shared by tRNA^{leu} and tRNA^{tyr}, is a crucial identity element has been shown by several biochemical studies (Normanly *et al.*, 1992; Sampson and Saks, 1993; Asahara *et al.*, 1994). SerRS is characterized by a remarkable 100 residue N-terminal domain which is folded into a 60 Å long, solvent exposed and flexible anti-parallel coiled-coil known as the helical arm. Crystallographic (Biou *et al.*, 1994; Cusack *et al.*, 1996a) and biochemical studies in which the helical arm is deleted (Borel *et al.*, 1994) show that this structural feature mediates the specific recognition of tRNA^{ser}. Another distinctive feature of the serine system is that the tRNA binds across the two subunits of the dimer, but still two tRNAs can bind simultaneously (Fig. 8). Upon tRNA binding the helical arm of the synthetase is stabilized in a new orientation and binds between the TΨC loop and the long variable arm of the tRNA. At the extreme end of the helical arm there is a hydrophobic platform upon which the tertiary base pair G19–C56 stacks (Biou *et al.*, 1994). Contacts with the tRNA long variable arm backbone extend until the sixth base pair, explaining the need for a minimum length of the tRNA variable arm. However the synthetase makes principally backbone contacts and few base-specific interactions and is mainly recognizing the unique shape of tRNA^{ser} (Asahara *et al.*, 1994). The characteristic shape of tRNA^{ser} is largely determined by bases 20A and 20B inserted into the D-loop which both play novel roles in tertiary interactions in the core of the tRNA (Biou *et al.*, 1994). In particular the base of G-20B is stacked against the first base pair of the long variable arm and thus defines the directional orientation of the latter, which is probably a distinguishing feature from the long variable arms of tRNA^{leu} and tRNA^{tyr}. In the original binary SerRS-tRNA^{ser} complex crystal structure (Biou *et al.*, 1994), the end of the acceptor stem was not ordered in the active site. However the ternary complex of SerRS-tRNA^{ser} with a non-hydrolysable seryl-adenylate analogue (Cusack *et al.*, 1996a) shows a much better ordered active site and the interactions inside the acceptor stem major groove made by the motif 2 loop of SerRS are visible (Fig. 9). The motif 2 loop (which is longer than that of AspRS) makes base contacts down to the fourth base pair of the acceptor stem which, however, are only weakly discriminatory, in agreement with biochemical results (Asahara *et al.*, 1994; Saks and Sampson, 1996). Interestingly, in the absence of tRNA, but in the presence of

ATP or seryl-adenylate (Belrhali *et al.*, 1995; Cusack *et al.*, 1996a) the motif 2 loop adopts a quite different ordered conformation. Upon tRNA binding a number of motif 2 loop residues previously found interacting with the ATP or adenylate now switch to participate in tRNA recognition. The conclusion from these results is that the functional binding of tRNA^{ser} to seryl-tRNA synthetase occurs in at least distinct two steps: firstly the initial recognition and docking which depends largely on the helical arm/tRNA long variable arm interaction on one subunit, and secondly, the correct positioning of the 3' end of the tRNA in the active site of the other subunit. The latter process depends critically on a conformational switch of the motif 2 loop after adenylate formation (Fig. 10). These results shed further light on the structural dynamics of the overall aminoacylation reaction in class II synthetases by revealing a mechanism which may promote an ordered passage through the activation and transfer steps.

Acknowledgements

We would like to thank Laurence Seignovert, Carmen-Berthet Colominas, Reuben Leberman and Michael Härtlein for agreement to include *T. thermophilus* asparaginyl-tRNA synthetase in Fig. 3. All figures except 1 and 9 have been prepared with MOLSCRIPT (Kraulis, 1991).

References

Åberg, A., Yaremchuk, A., Tukalo, M., Rasmussen, B. and Cusack, S. (1997). *Biochem.* In press.

Arnez, J. G. and Moras, D. (1994) In *RNA–Protein Interactions* (K. Nagai and I. W. Mattai, eds), pp. 52–81. IRL Press, Oxford.

Arnez, J. G., Harris, D. C., Mitschler, A., Rees, B., Francklyn, C. S. and Moras D. (1995). *EMBO J.* **14**, 4143–4155.

Asahara, H., Himeno, H., Tamura, K., Nameki, N., Hasegawa, T. and Shimizu, M. (1994). *J. Mol. Biol.* **236**, 738–748.

Belrhali, H., Yaremchuk, A., Tukalo, M., Berthet-Colominas, C., Rasmussen, B., Bösecke, P., Diat, O. and Cusack, S. (1995). *Structure* **3**, 341–352.

Biou, V., Yaremchuk, A., Tukalo, M. and Cusack, S. (1994). *Science* **263**, 1404–1410.

Bochkarev, A., Pfuetzner, R. A., Edwards, A. M. and Frappier, L. (1997). *Nature* **385**, 176–181.

Borel, F., Vincent, C., Leberman, R. and Härtlein, M. (1994). *Nucleic Acids. Res.* **22**, 2963–2969.

Brunie, S., Zelwer, C. and Risler, J-L. (1990). *J. Mol. Biol.* **216**, 411–424.

Bycroft, M., Hubbard, T. J. P., Proctor, M., Freund, S. M. V. and Murzin, A. G. (1997). *Cell* **88**, 235–242.

Cavarelli, J., Rees, B., Ruff, M., Thierry, J.-C. and Moras, D. (1993). *Nature* **362**, 181–184.

Cavarelli, J., Eriani, G., Rees, B., Ruff, M., Boeglin, M., Mitschler, A., Martin, F., Gangloff, J., Thierry, J.-C. and Moras, D. (1994). *EMBO J.* **13**, 327–337.

Cusack, S. (1995). *Nature Struc. Biol.* **2**, 824–831.

Cusack, S., Berthet-Colominas, C., Härtlein, M., Nassar, N. and Leberman, R. (1990). *Nature* **347**, 249–255.

Cusack, S., Härtlein, M. and Leberman, R. (1991). *Nucleic Acids. Res.* **19**, 3489–3498.

Cusack, S., Yaremchuk, A. and Tukalo, M. (1996a). *EMBO J.* **15**, 2834-2842.
Cusack, S., Yaremchuk, A. and Tukalo, M. (1996b). *EMBO J.* **15**, 6321-6334.
Delarue, M., Poterszman, A., Nikonov, S., Garber M., Moras D., Thierry J. C. (1994). *EMBO J.* **13**, 3219-3229.
Delarue, M. and Moras, D. (1993). *Bioessays* **15**, 675-687.
Doublié, S., Bricogne, G., Gilmore, C. and Carter, C. W., Jr. (1995) *Structure* **3**, 17-31.
Eriani, G., Delarue, M., Poch, O., Gangloff, J. and Moras, D. (1990). *Nature* **347**, 203-206.
Fraser, T. H. and Rich, A. (1975). *Proc. Natl. Acad. Sci. USA* **72**, 3044-3048.
Goldgur, Y., Mosyak, L., Reshetnikova, L., Ankilova, V., Lavrik, O., Khodyreva, S. and Safro, M. (1997). *Structure* **5**, 59-68.
Himeno H., Hasegawa, T., Ueda, T., Watanabe, K., Miura, K. and Shimizu, M. (1989). *Nucleic Acids Res.* **19**, 7855-7863.
Hou, Y.-M. and Schimmel, P. (1988). *Nature* **333**, 140-145.
Kraulis, P. J. (1991). *J. Appl. Cryst.* **24**, 946-950.
Lin, L., Hale, S. P. and Schimmel, P. (1996). *Nature* **384**, 33-34.
Logan, D. T., MazaUc, M-H., Kern, D. and Moras, D. (1995). *EMBO J.* **14**, 4156-4167.
Mosyak, L., Reshetnikova, L., Goldgur, Y., Delarue, M. and Safro, M. G. (1995). *Nature Struc. Biol.* **2**, 537-547.
Murzin, A. G. (1993). *EMBO J.* **12**, 861-867.
Normanly, J., Ogden, R. C., Horvath, S. J., Abelson, J. (1986). *Nature* **321**, 213-219.
Normanly, J., Ollick, T., Abelson, J. (1992). *Proc. Natl. Acad. Sci. USA* **89**, 5680-5684
Nureki, O. *et al.* (1995). *Science* **267**, 1958-1965.
Onesti, S., Miller, A. D. and Brick, P. (1995). *Structure* **3**, 163-176.
Rould, M. A., Perona, J. J., Söll, D. and Steitz, T. A. (1989). *Science* **246**, 1135-1142.
Rould, M. A., Perona, J. J. and Steitz, T. A. (1991). *Nature* **352**, 213-218.
Ruff, M. *et al.* (1991). *Science* **252**, 1682-1689.
Saks, M. E., Sampson, J. R. and Abelson, J. N. (1994). *Science* **263**, 191-197.
Saks, M. E. and Sampson, J. R. (1996). *EMBO J.* **15**, 2843-2849.
Sampson, J. R. and Saks, M. E. (1993). *Nucleic Acids Res.* **21**, 4467-4475.
Sprinzl, M. and Cramer, F. (1975). *Proc. Natl. Acad. Sci. USA.* **72**, 3049-3053.
Vincent, C., Borel, F., Willison, J. C., Leberman, R. and Hartlein, M. (1995). *Nucleic Acids Res.* **23**, 1113-1118.

5

RNA−Protein Recognition during RNA Processing and Maturation*

GABRIELE VARANI

MRC Laboratory of Molecular Biology, Hills Road, Cambridge CB2 2QH, UK

Abstract

Gene expression in eukaryotes is regulated after transcription through messenger RNA (mRNA) stability, transport and localization and through the excision of non-coding regions (splicing). During these maturation events, mRNAs are complexed in ribonucleoprotein particles with proteins that recognize specific RNA sequences to affect different regulatory steps. The assembly of large ribonucleoproteins perfoming RNA splicing (spliceosome) and translation (ribosome) also depends on recognition of spliceosomal small nuclear RNAs (snRNAs) and ribosomal RNA, respectively, by constitutive and auxiliary protein factors. Understanding the molecular basis of RNA−protein recognition is necessary to understand these regulatory events and to learn how to exogeneously regulate gene-expression at the post-transcriptional stage. This review describes our current knowledge of the molecular basis of RNA−protein recognition events that occur during post-transcriptional RNA processing.

1 Introduction

Biochemical and thermodynamic data on RNA−protein complexes reveal that intermolecular interactions occur with the appropriate affinity to allow binding and regulation at intra-cellular concentrations of RNA and protein components. For example, aminoacyl tRNA-synthetase (aaRS) enzymes bind tRNA substrates with modest affinity (bimolecular dissociation constant $K_d \approx 10^{-6}$ M). AaRS enzymes catalyse the covalent attachment of the 20 amino acids to the $3'$-end of transfer RNA (tRNA). The modest affinity is necessary to facilitate substrate

---

*Figures 6 and 7 for this chapter appear in the colour plate section between p. 144 and p. 145

THE MANY FACES OF RNA
ISBN 0-12-233210-5

release and improve catalytic efficiency. By contrast, constitutive components of stable ribonucleoproteins (such as the human U1A and U1 70 K proteins, both are components of the pre-mRNA splicing machinery) bind cognate RNAs much more tightly ($K_d < 10^{-9}$ M). In addition to absolute binding energies (or dissociation constants), intermolecular interactions are also characterized by differences in binding free energy for different substrates. Many RNA-binding proteins bind any RNA with low affinity, regardless of its sequence and structure. However, biological function requires discrimination of cognate RNAs from non-cognate sequences, which are present in large excess in the cell nucleus. Differences in binding energy between cognate and non-cognate RNAs define the specificity of an interaction.

Understanding the molecular basis of RNA–protein recognition requires a rationalization of the physical origin of the absolute binding energy (affinity) and of relative energy differences for different substrates (specificity). The analysis of the relatively few existing structures of RNA–protein complexes determined at atomic resolution reveals that affinity and specificity do not depend in a simple way on simple physical chemical properties, such as protein charge or size of intermolecular interface area. For example, tRNA–synthetase complexes have very large interface areas (≈ 3000 Å^2) compared to the much tighter complexes involving the human U1A protein (surface area ≈ 1300 Å^2).

2   RNA recognition differs from DNA recognition

In contrast to the few structures of RNA–protein complexes, over 150 existing structures of DNA–protein complexes define an important paradigm in intermolecular recognition. Very often, DNA–protein recognition occurs by insertion of an α-helix into the major groove of double-stranded DNA (Steitz, 1990). Specific sequences are then recognized by direct readout of hydrogen bond donors and acceptors on the DNA bases and by indirect readout of sequence-dependent DNA conformational features through van der Waals interactions and through electrostatic interactions with the negatively charged phosphodiester backbone. These principles do not apply to RNA recognition. The major groove of double helical RNA is too narrow and deep (Fig. 1) to allow the insertion of protein secondary structural elements (α-helices or β-sheet), as would be necessary for the efficient discrimination of different base pairs. The RNA minor groove is wide and easily accessible (Fig. 1). However, there is insufficient diversity of functional groups in the minor groove to allow effective sequence discrimination (Seeman et al., 1976): all the base pairs are very similar to each other when viewed from the minor groove. Although some DNA-binding proteins (namely HMG-box proteins (Love et al., 1995; Werner et al., 1995) and the TATA-box binding protein (Kim et al., 1993; Kim et al., 1993) recognize the DNA minor groove, only limited sequence discrimination can generally be provided when recognition occurs in the minor groove.

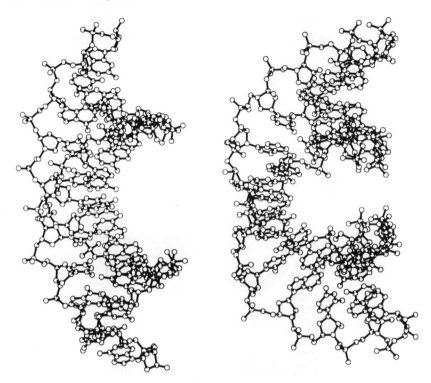

Fig. 1 DNA and RNA double helices present different structures. The wide major groove of B-form DNA (left) sharply contrasts with the narrow and deep major groove of A-form RNA (right).

Given these considerations, it is not surprising that there is no example of sequence-specific recognition of RNA double helices. Sequence-specific RNA-binding proteins recognize instead single-stranded regions and hairpin loops where the functional groups on the bases are accessible for recognition (Fig. 2). Double-helical regions are recognized only when distortions in the double helix generated by internal loops (Battiste *et al.*, 1996; Ye *et al.*, 1996) or bulges (Puglisi *et al.*, 1995; Ye *et al.*, 1995) or at the helix termini (Rould *et al.*, 1989; Rould *et al.*, 1991) allow access to base functionalities in the major groove.

Another major difference between DNA and RNA is the diversity of RNA structure. To a first approximation, all DNA double helices adopt the same three-dimensional structure (Fig. 1). Although sequence-dependent conformational features have been repeatedly observed and are undoubtedly critical for protein-DNA recognition, these features constitute relatively minor distortions from the canonical double helical structure. The diversity of RNA functions requires an equally diverse array of tertiary structural features. This diversity allows the recognition of unique three-dimensional surfaces and charge distributions of different RNAs and defines an enormous variety of specific interactions.

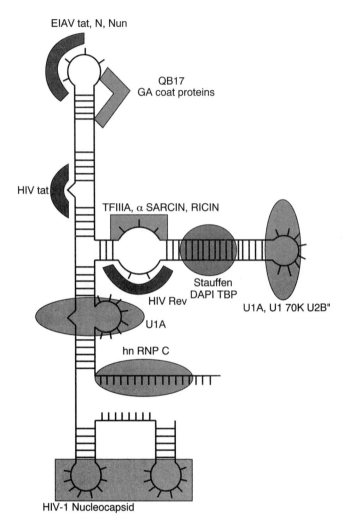

Fig. 2   RNA-binding proteins do not target double-stranded RNA in a sequence specific manner, but recognize instead single-stranded regions (hnRNP C) or sites of local distortions induced in double helical regions by RNA hairpins (U1A, U1 70K, EIAV Tat, N, ...), bulges (HIV Tat) or internal loops (U1A, HIV Rev, TFIIIA, ...).

3   αβ protein domains

Proteins involved in RNA metabolism from a wide spectrum of organisms have a highly modular structure. RNA-binding modules of 60–90 amino acids are often juxtaposed to auxiliary domains that perform additional functions (Biamonti and Riva, 1994; Burd and Dreyfuss, 1994; Nagai, 1996), such as mediating protein–protein interactions or regulating RNA-binding activity through phosphorylation. The three most common RNA-binding modules are the

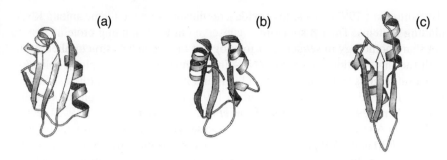

Fig. 3 Three-dimensional structure of the three most common RNA-binding folds: (a) RNP domain (Nagai *et al.*, 1990); (b) KH domain (Musco *et al.*, 1996) and (c) dsRBD domain (Bycroft *et al.*, 1995; Kharrat *et al.*, 1995).

ribonucleoprotein domain (RNP, often referred to as RRM or RNA recognition motif), K-homology (KH) domain and double-stranded RNA binding (dsRBD) domain. Each of these motifs are characterized by a compact, globular structure (Fig. 3) and constitute independent structural domains and RNA-binding motifs (Nagai, 1996).

The dsRBD domain is a general, non-sequence-specific double-stranded RNA-binding module (St Johnston *et al.*, 1992; Bycroft *et al.*, 1995; Kharrat *et al.*, 1995). Isolated domains bind double-stranded RNA without sequence specificity (Clarke and Mathews, 1995), but multiple dsRBD domains may recognize certain RNA structures specifically (Ferrandon *et al.*, 1994; Clarke and Mathews, 1995). KH domains appear to be non-sequence-specific single-stranded RNA-binding proteins associated with important metabolic functions. Depressed expression of the KH-protein FMR1 or a single amino acid substitution that unfolds a KH domain within this protein (Musco *et al.*, 1996), lead to fragile-X syndrome (Ashley *et al.*, 1993; Siomi *et al.*, 1993), the most common cause of genetically inherited mental retardation in humans. The RNP domain is one of the most common eukaryotic protein folds, comprising over 600 sequences. RNP proteins specifically recognize single-stranded and highly structured RNA elements. The RNP motif is identified by two highly conserved amino acid sequences (RNP-1 and RNP-2) (Query *et al.*, 1989) located in the two central strands of a 4-stranded antiparallel β-sheet (Nagai *et al.*, 1990; Hoffman *et al.*, 1991). Sequences related to RNP-1 and RNP-2 have been identified in prokaryotic RNA-binding proteins, including the bacteriophage T4 translational repressor RegA (Kang *et al.*, 1995), initiation factor IF3 (Biou *et al.*, 1995; Garcia *et al.*, 1995) and termination factor *Rho* (Martinez *et al.*, 1996 a,b). RNP (Nagai *et al.*, 1990), KH (Musco *et al.*, 1996), and dsRBD (Bycroft *et al.*, 1995; Kharrat *et al.*, 1995) are $\alpha\beta$ proteins. In all three cases, an antiparallel β-sheet on one face of the protein is packed by a hydrophobic core against an α-helical face (Fig. 3). Although all α-helical RNA-binding proteins have recently been identified (Berglund *et al.*, 1997; Markus *et al.*, 1997; Xing *et al.*, 1997), $\alpha\beta$ domains and all-β structures

(Bycroft *et al*., 1997) appear to provide a dominant structural theme among RNA-binding proteins. The $\alpha\beta$ structure is conserved in RNA-binding proteins that do not share homology in sequence or topology of the secondary structural elements with the three major eukaryotic RNA-binding motifs. For example, a split $\beta\alpha\beta$ structure reminiscent of the RNP-fold is found in six ribosomal proteins (S6, L1, L6, L9, L12 and L30) (Liljas and Garber, 1995), in translational elongation (Liljas and Garber, 1995) and initiation (Biou *et al*., 1995; Garcia *et al*., 1995) factors and in RegA (Kang *et al*., 1995). In some cases (for example ribosomal proteins L12 and L30 and RNP proteins, or dsRBD and ribosomal protein S5) the similar secondary structure topology indicates a common evolutionary origin. However, $\alpha\beta$ RNA-binding proteins have different topology of the secondary structure elements. The RNP domain has a repeated $\beta\alpha\beta$ arrangement (Nagai *et al*., 1990; Hoffman *et al*., 1991), dsRBDs have $\alpha\beta\beta\beta\alpha$ topology (Bycroft *et al*., 1995; Kharrat *et al*., 1995), and KH proteins have $\beta\alpha\alpha\beta\beta\alpha$ fold (Musco *et al*., 1996). Topological differences and low sequence homology indicate that the different domains represent distinct, convergent evolutionary solutions towards a common structure for RNA recognition.

4   The RNP domain paradigm of RNA recognition

Hundreds of proteins involved in RNA metabolism recognize substrates widely diverse in sequence and structure via RNP domains. The identity of RNA substrates of RNP proteins is often defined both by RNA structural features and by recognition of single-stranded nucleotides. Multiple RNP domains are required for recognition of single-stranded RNA sequences with high affinity and specificity (Shamoo *et al*., 1994; Kanaar *et al*., 1995; Tacke and Manley, 1995). However, single RNP domains can recognize 5–10 single-stranded RNA bases with high affinity and specificity when the nucleotides are presented in a defined RNA structural context. For example, the human U1A protein recognizes seven single-stranded nucleotides in the context of a hairpin or internal loop with very tight binding constant $K_d \approx 10^{-10}$–10^{-11} M) (Hall and Stump, 1992; Nagai *et al*., 1990). Crystallographic (Oubridge *et al*., 1994) and NMR (Allain *et al*., 1996) structures of the complexes of the human U1A protein with two distinct RNA substrates have revealed important aspects of the molecular basis of RNP–RNA recognition. These complexes provide the only existing high-resolution structures of RNA–protein complexes from the RNA processing and maturation machinery.

The N-terminal RNP domain of U1A binds a hairpin loop during pre-mRNA splicing (Scherly *et al*., 1990; Scherly *et al*., 1991) and two related internal loops (Fig. 4) during autoregulation of polyadenylation (Boelens *et al*., 1993; van Gelder *et al*., 1993; Gunderson *et al*., 1994). Hairpin and internal loop substrates present similar single-stranded sequences in completely different structural contexts. In both the crystal structure of the U1A-hairpin complex (Oubridge *et al*., 1994) and the solution structure of the internal loop complex (Allain *et al*.,

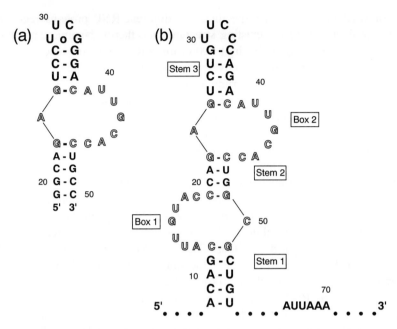

Fig. 4 The human U1A protein recognizes (a) a hairpin loop during splicing and (b) a repeated internal loop during regulation of polyadenylation.

1996), bases within the single-stranded loops are splayed out across the surface of the antiparallel β-sheet and are all involved in intra- or intermolecular base stacking interactions. Most hydrogen bond donors and acceptors from the single-stranded bases are recognized by an extensive hydrogen bonding network with protein residues from the β-sheet and from three loops connecting the secondary structural elements. Perhaps surprisingly, a large number of intermolecular contacts involve protein main chain hydrogen bond donors and acceptors. The RNA backbone follows a region of positive electrostatic potential in the protein to provide favourable electrostatic interactions (Oubridge *et al.*, 1994; Nagai *et al.*, 1995).

5 RNP proteins recognize both the identity of single-stranded nucleotides and the unique shape and charge distribution of the substrate RNA

The stereochemical complementarity between the β-sheet of U1A and its RNA substrates allows the formation of a very large number of intermolecular interactions (hydrogen bonding, ring stacking, van der Waals and electrostatic contacts). These interactions provide (at least in a qualitative sense) a rationale for the very low dissociation constants observed for these complexes ($K_d \approx 10^{-11}$ M) (Hall and Stump, 1992; van Gelder *et al.*, 1993). However, the equally compelling issue of what determines the specificity of different RNP proteins requires a careful consideration of the existing structures and thermodynamic data and of the RNP sequence database. Each of 600 or more RNP domains binds a distinct RNA target but does not bind effectively other RNAs. What is the structural basis for this remarkable ability to discriminate different RNA substrates?

A superficial examination of the U1A–RNA structures would lead to the suggestion that recognition of the exposed single-stranded nucleotides is responsible for the specificity of the interaction. In fact, each of the seven single-stranded nucleotides is recognized by a very extensive network of interactions with protein residues and elegant *in vitro* genetic experiments have demonstrated that each of these single-stranded nucleotides is required for optimal binding to the U1A protein (Harper *et al.*, 1992). However, the U1A-related U2B″ protein does not bind at all the U1A substrates (Scherly *et al.*, 1990), despite being 80% identical to U1A and sharing almost every amino acid involved in recognition of the seven single-stranded nucleotides. Thus, two closely related proteins with a nearly identical potential to recognize the same single-stranded loop have very different substrate specificity! This powerful argument against a dominant role for single-stranded RNA recognition is reinforced by the observation that the surface of the β-sheet of RNP domains is so highly conserved among different proteins that it is difficult to see how discrimination can be achieved. To reinforce this observation, we compared the sequence of amino acids involved in single-stranded nucleotide recognition from RNP proteins with completely different specificity (U1A, poly-A binding protein that binds poly-A and proteins such as *Drosophila sxl* or human U2AF that bind pyrimidine-rich sequences). We found that only six hydrogen bonding interactions from four unique amino acids located on the β-sheet surface differ between these proteins. Thus, although recognition of single-stranded nucleotides is an important determinant of U1A specificity, it is not sufficient for intermolecular discrimination.

Important clues towards understanding the origin of the binding specificity of U1A are provided by two observations. Firstly, although the identity of the seven single-stranded nucleotides is recognized specifically even in the absence of RNA secondary structure (Tsai *et al.*, 1991) the binding constant is reduced 100 000 fold when the RNA secondary structure is destroyed. This reduction brings the binding constant close to the level observed for binding to RNA of random sequence (Hall, 1994). Thus, RNA structure appears to be an important determinant of the specificity of binding. Secondly, the analysis of the RNP protein sequence database reveals that the sites of greatest genetic diversity are the loops connecting the secondary structural elements (Fig. 5). Before the structure of any complex was available, it had already been suggested that the β-sheet surface of the RNP domain represents a generic RNA-binding platform, to which identity elements are provided by these variable loops and by sequences N- and C-terminal to the domain itself (Kenan *et al.*, 1991; Burd and Dreyfuss, 1994). What is the structural role of these variable loops in U1A–RNA recognition?

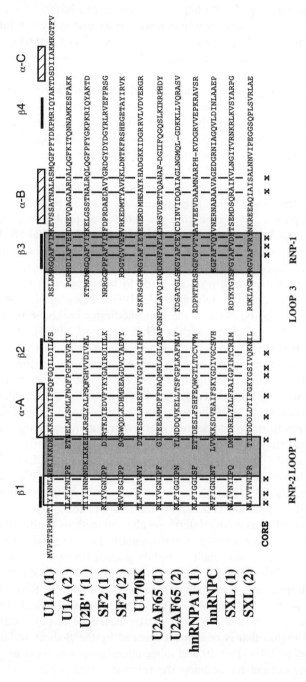

Fig. 5 Alignment of amino acid sequences of selected RNP proteins; this alignment is based on the analysis of Birney and coworkers (Birney *et al.*, 1993). The variable loops and the two highly conserved RNP-1 and RNP-2 sequences within the central strands of the antiparallel β-sheet are highlighted.

The region C-terminal to the RNP domain (residues 88–98) is critical for U1A–RNA binding (Fig. 5). Proteins truncated at residues 92 have severely reduced binding constant, and even a construct truncated at Lys 96 binds RNA more weakly than the full-length construct (Jessen *et al.*, 1991). The loop connecting the fourth strand of the antiparallel sheet to the C-terminal Helix C (residues 88–92) forms many interactions with RNA bases. Surprisingly, every contact involves main chain amide and carbonyl functionalities rather than protein side chains. Specificity is achieved by a distinctive location of the backbone of Helix C. In the free U1A protein, Helix C lies across the surface of the β-sheet and forms a small hydrophobic patch involving the hydrophobic surface of this helix (residues Ile 93, Ile 94 and Met 97) and several heteroaromatic and hydrophobic residues on the β-sheet surface (Avis *et al.*, 1996). These residues become exposed for RNA binding by a sharp conformational rearrangement centred around residues 88–90. This movement is favoured by the formation of a new hydrophobic patch involving the same hydrophobic face of Helix C and different parts of the protein, namely His 10, Leu 41 and Val 58 (Fig. 6a).

The loop connecting the first strand of the β-sheet and Helix A (loop 1) is characterized in U1A by the presence of several Lys residues (Fig. 5). Substitution of some of these basic residues leads to large losses in binding energy for the U1A-hairpin complex (Nagai *et al.*, 1990; Jessen *et al.*, 1991). These residues also appear to be close to the phosphodiester backbone in the hairpin complex determined crystallographically (Oubridge *et al.*, 1994), in the internal loop complex they are in close contact with the phosphodiester backbone of one of the helical stems (together with Arg 47 from loop 3) (Fig. 6b). These electrostatic interactions are likely to provide large contributions to the overall free energy of binding, but also contribute to the ability of U1A to discriminate different RNA substrates. Since these residues are unique to U1A among all RNP proteins, these electrostatic interactions must contribute to specificity as well.

The loop connecting the second and third strand of the β-sheet (loop 3, residues 45–52) is in intimate contact with the RNA; residues from this loop form many intermolecular hydrogen bonding and hydrophobic contacts. An elegant analysis of the requirements for U1A–RNA recognition using the phage display technology led to the surprising conclusion that many residues could be mutated, even those involved in extensive intermolecular interactions (Laird-Offringa and Belasco, 1995). However, Leu 49 was particularly important for intermolecular recognition. The Leu 49 side chain is tightly packed between a G–C base pair and a well-stacked Adenine at the stem-loop junction, and every single-side chain carbon and hydrogen atom is in van der Waals contacts with the RNA. This interaction appears to play the role of a 'pivot' connecting the helical stem (where electrostatic recognition of the phosphodiester backbone is observed) with the single-stranded region that is recognized instead by the β-sheet surface and C-terminal regions of U1A. Leu 49 and other interactions involving neighbouring residues may be critical for defining the relative spatial location of these two regions of intermolecular contact.

Recognition of the unique shape and charge distribution of the RNA internal loop structure by basic amino acids found only in U1A (Fig. 6b) is clearly a critical step in intermolecular discrimination. However, it is difficult to see how this mechanism could be effective for proteins that recognize specifically single-stranded RNA sequences that lack distinctive structural features. Interestingly, in all known cases, single-stranded RNA recognition requires more than a single RNP domain. Tight, highly specific binding occurs only when multiple RNP domains are present: isolated domains generally bind RNA more weakly and less specifically (Kanaar et al., 1995; Tacke and Manley, 1995). Interdomain interactions appear to play a critical structural role in the recognition of single-stranded RNA sequences by RNP proteins. The determination of the structure of specific complexes between single-stranded RNA and multiple-domain RNP proteins is the next challenge in understanding RNP-RNA recognition; such studies are underway in a number of laboratories.

6 Rigid docking and adaptive binding

The analysis of the structure of the RNA-U1A complexes has provided important clues towards understanding the stereochemical basis of RNP-RNA recognition. However, the comparison with the unbound RNA internal loop (Gubser and Varani, 1996) and protein (Avis et al., 1996) structures (Fig. 7) reveals a further level of complexity in intermolecular recognition. Binding does not occur by rigid docking of pre-determined RNA and protein structures, nor by pure induced fit. Rather, part of the intermolecular interface appears to be pre-formed, but five single-stranded nucleotides are recognized by induced fit. A first set of inter-molecular contacts involve rigid docking between the protein and the RNA double helical regions and three well ordered single-stranded nucleotides. This rigid interaction involves the loops (β2-β3 loops and β4-Helix C) that represent the identity elements of U1A and other RNP proteins. Protein binding then orders five single-stranded nucleotides (disordered in the free RNA structure) against the β-sheet surface through intermolecular stacking interactions with three very highly conserved aromatic amino acids. RNP1- and RNP2-like sequences with exposed aromatic residues are found in other RNA-binding (Biou et al., 1995; Garcia et al., 1995; Kang et al., 1995; Martinez et al., 1996 a,b) and single-stranded DNA-binding (Schindelin et al., 1993; Schnuchel et al., 1993) proteins. Inter-molecular stacking interactions with exposed aromatic side chains are common to $\alpha\beta$ proteins (Oubridge et al., 1994; Allain et al., 1996), tRNA synthetases (Rould et al., 1991; Caverelli et al., 1993) and viral coat proteins (Valegård et al., 1994) and may be duplicated in ribosomal proteins (Liljas and Garber, 1995) and translational regulators (Biou et al., 1995; Garcia et al., 1995; Kang et al., 1995; Martinez et al., 1996 a,b). Recent thermodynamic results suggest that these interactions provide large amounts of binding free energy (LeCuyer et al., 1996).

7 Conformational flexibility at RNA–protein interfaces

Consideration of the entropically costly disorder–order transition in the single-stranded RNA loops makes the very tight binding of the U1A protein ($K_d \approx 10^{-11}$ M) even more remarkable. Changes in the distribution of molecular vibrations when exposed amino acid side chains become ordered at intermolecular interfaces contribute 15–25 kcal mol^{-1} to increase the free energy of protein–protein and protein–DNA interactions (Akke et al., 1993). However, the NMR relaxation properties of protein side chains in DNA–protein and protein–protein complexes indicate that motion is less restricted at the intermolecular interface than in the highly ordered hydrophobic core (Berglund et al., 1995; Kay et al., 1996; Slijper et al., 1997), suggesting that protein–nucleic acids interfaces may not be rigidly ordered.

Several observations suggest that these considerations apply to the U1A complexes as well. Firstly, the Arg 52 side chain in U1A is involved in five hydrogen bonding interactions in the crystal structure (Oubridge et al., 1994), but can be mutated to Lys (which cannot form those same hydrogen bonds) without significant increase in the free energy of binding ($\Delta\Delta G < 0.5$ kcal mol^{-1}) (Nagai et al., 1990; Jessen et al., 1991). The NMR data suggest a less ordered conformation, where each hydrogen bond is only present part of the time (Allain et al., 1996). Secondly, a C $\rightarrow$ G substitution only leads to a 10-fold decrease in binding constant (Hall and Stump, 1992). In both the NMR and X-ray structures (Oubridge et al., 1994; Allain et al., 1996), that cytosine is deeply buried at the intermolecular interface preventing fitting of a larger guanosine, and every cytosine functionality is recognized by the protein. An explanation consistent with several NMR observations is that the interface is flexible enough to accommodate either base through energetically inexpensive local conformational adjustments.

Studies of protein dynamics at intermolecular interfaces suggest that a fine balance between rigidity and flexibility may provide a compromise between complete specificity (at large entropic cost) and complete lack of selectivity. Characterization of the flexibility of the U1A intermolecular interface through NMR relaxation measurements probing the ns–ps time scale would address these issues.

In contrast to the situation observed with induced fit, it is more difficult to reorganize the intermolecular interface when the interaction occurs by rigid fit between highly ordered regions of the protein and RNA. Thus, it is costly to mutate the tightly packed Leu 49 in U1A at the junction between RNA helices and loops (Laird-Offringa and Belasco, 1995; Allain et al., 1996). The highly conserved Arg 52 nearby can only be mutated to Lys, which has similar size and positive charge (Nagai et al., 1990; Jessen et al., 1991). It has already been mentioned that disruption of the RNA secondary structure reduces binding 100 000 fold (Tsai et al., 1991). The pre-formed RNA secondary structures provide a structural counterpart to the β-sheet of the protein in reducing the entropic costs of RNA folding (Nagai et al., 1995) and may provide large amounts of binding free energy through essential electrostatic interactions.

8 Do RNA and protein conformational rearrangements contribute to specificity?

Characteristically large heat capacity changes ($\Delta C_p \leq 0$) observed in DNA–protein complexes have been convincingly attributed to local protein folding on DNA binding to create key parts of the intermolecular interface (Spolar and Record, 1994). Protein folding generally dominates the thermodynamic contribution from changes in surface area upon DNA-binding: proteins have greater conformational adaptability compared to DNA. Thermodynamic data show that large heat capacity changes also occur upon U1A–RNA binding (Hall and Stump, 1992; Hall, 1994). Formation of the U1A–RNA interface through induced fit depends on the identity of amino acid side chains exposed on the β-sheet surface and in neighbouring loops, as well as on the identity of the single-stranded nucleotides. Both the driving force for the conformational change (the binding free energy) and the driven process (RNA folding) depend on RNA and protein sequences, providing a potentially crucial step in intermolecular discrimination: binding energy could be dissipated to favour the conformational change that would enhance discrimination. Since changes in sequence may critically affect the geometry of the interface far away from the site where the mutation occurs, non-additive energetics of recognition are expected. In fact, mutations in U1A or in its RNA targets cannot be easily interpreted from the loss of intermolecular contacts observed in NMR and X-ray structures (Nagai et al., 1990; Jessen et al., 1991; Hall and Stump, 1992; Hall, 1994; Policarpou-Schwartz et al., 1996).

9 Functional implications

Direct evidence for the important functional role of protein-induced RNA conformational rearrangements is provided by self-splicing autocatalytic RNAs (Mohr et al., 1992; Mohr et al., 1994; Weeks and Cech, 1995; Weeks and Cech, 1996). Group I introns are RNA enzymes that often require protein cofactors for catalysis in vivo. Binding of a fungal tyrosyl-tRNA synthetase (Cyt-18) to the catalytic core of a mitochondrial ribosomal RNA intron folds the pre-existing RNA secondary structure into the catalytically active tertiary structure (Mohr et al., 1992; Mohr et al., 1994). The yeast CBP2 protein enhances splicing of a group I mitochondrial intron by inducing docking of the substrate into the enzyme active core (Weeks and Cech, 1995; Weeks and Cech, 1996). In both cases, protein binding induces higher-order tertiary RNA structures required for catalytic function. Similarly, ribosomal proteins induce formation or stabilization of a compact tertiary structure which is largely absent in the naked ribosomal RNA. Protein subunits of RNA-based enzymes perform essential auxiliary functions, such as stabilizing or modulating RNA structures and allowing regulation, through protein-induced changes in RNA conformation.

10   Conclusions and perspectives

Recent crystallographic and solution NMR structures of RNA–protein complexes have revealed principles of molecular recognition by RNA-binding proteins. These structures demonstrate different ways in which protein β-sheets provide large surfaces for extensive interactions with RNA bases exposed in single-stranded regions. Exposed β-sheet surfaces are so common in RNA recognition to suggest a role as dominant as that of α-helices in DNA recognition. This dominance probably originates from the stereochemical complementarity with RNA structure due to the natural right-handedness and concavity of antiparallel β-sheets (Jessen *et al.*, 1991; Howe *et al.*, 1994; Liljas and Garber, 1995). A balance of induced fit and shape selectivity through rigid fit is common to tRNA-synthetase complexes and to the recognition of highly structured RNAs by RNP proteins. Formation of RNA–protein complexes is clearly a highly dynamic process: RNA structure directs protein binding, which in turn modulates the RNA conformation to create a unique intermolecular interface.

   The structural and functional diversity of RNA potentially define a large variety of recognition mechanisms. However, atomic resolution structural information is only available for a handful of RNA–protein complexes, excluding any ribosomal component, KH or dsRBD domains. The next several years will undoubtedly see exciting progress in understanding the thermodynamic and structural basis of recognition. Besides addressing fundamental questions in intermolecular recognition, these studies will underpin the search for compounds that interfere with critical RNA–protein recognition events to down-regulate gene expression and prevent replication of pathogenic viruses and bacteria (Gait and Karn, 1995).

Acknowledgements

It is a pleasure to thank Charles Gubser and Luca Varani for preparation of illustrations; Dr Annalisa Pastore (EMBL, Heidelberg) for sharing the coordinates of KH and dsRBD proteins determined in her laboratory.

References

Akke, M., Brüschwiler, R. and Palmer, A. G. I. (1993). *J. Am. Chem. Soc.* **115**, 9832–9833.
Allain, F.-H. T., Gubser, C. C., Howe, P. W. A., Nagai, K., Neuhaus, D. and Varani, G. (1996). *Nature* **380**, 646–650.
Ashley, C. T. J., Wilkinson, K. D., Reines, D. and Warren, S. T. (1993). *Science* **262**, 563–566.
Avis, J., Allain, F. H.-T., Howe, P. W. A., Varani, G., Neuhaus, D. and Nagai, K. (1996). *J. Mol. Biol.* **257**, 398–411.
Battiste, J. L., Mao, H., Rao, N. S., Tan, R., Muhandiram, D. R., Kay, L. E., Frankel, A. D. and Williamson, J. R. (1996). *Science* **273**, 1547–1551.
Berglund, H., Baumann, H., Knapp, S., Ladenstein, R. and Härd, T. (1995). *J. Am. Chem. Soc.* **117**, 12883–12884.

Berglund, H., Rak, A., Serganov, A., Garber, M. and Härd, T. (1997). *Nature Struc. Biol.* **4**, 20-23.

Biamonti, G. and Riva, S. (1994). *FEBS Letters* **340**, 1-8.

Biou, V., Shu, F. and Ramakrishnan, V. (1995). *EMBO J.* **14**, 4056-4064.

Birney, E., Kumar, S. and Krainer, A. R. (1993). *Nucleic Acids Res.* **21**, 5803-5816.

Boelens, W. C., Jansen, E. J. R., van Venrooij, W. J., Stripecke, R., Mattaj, I. W. and Gunderson, S. I. (1993). *Cell* **72**, 881-892.

Burd, C. G. and Dreyfuss, G. (1994). *Science* **265**, 615-621.

Bycroft, M., Grünert, S., Murzin, A. G., Proctor, M. and St Johnston, D. (1995). *EMBO J.* **14**, 3563-3571.

Bycroft, M., Hubbard, T. J. P., Proctor, M., Freund, S. M. V. and Murzin, A. G. (1997). *Cell* **88**, 235-242.

Caverelli, J., Rees, B., Ruff, M., Thierry, J.-C. and Moras, D. (1993). *Nature* **362**, 181-184.

Clarke, P. A. and Mathews, M. B. (1995). *RNA* **1**, 7-20.

Ferrandon, D., Elphick, L., Nüsslein-Volhard, C. and St Johnston, D. (1994). *Cell* **79**, 1221-1232.

Gait, M. J. and Karn, J. (1995). *TIBS* **13**, 430-438.

Garcia, C., Fortier, P.-L., Blanquet, S., Lallemand, J.-Y. and Dardel, F. (1995). *J. Mol. Biol.* **254**, 247-259.

Gubser, C. C. G. and Varani, G. (1996). *Biochemistry* **35**, 2253-2267.

Gunderson, S. I., Beyer, K., Martin, G., Keller, W., Boelens, W. C. and Mattaj, I. W. (1994). *Cell* **76**, 531-541.

Hall, K. B. (1994). *Biochemistry* **33**, 10076-10088.

Hall, K. B. and Stump, W. T. (1992). *Nucleic Acids Res.* **20**, 4283-4290.

Harper, D. S., Fresco, L. D. and Keene, J. D. (1992). *Nucleic Acids Res.* **20**, 3645-3650.

Hoffman, D. W., Query, C. C., Golden, B. L., White, S. W. and Keene, J. D. (1991). *Proc. Natl. Acad. Sci. USA* **83**, 2495-2499.

Howe, P. W. A., Nagai, K., Neuhaus, D. and Varani, G. (1994). *EMBO J.* **13**, 3873-3881.

Jessen, T. H., Oubridge, C., Teo, C. H., Pritchard, C. and Nagai, K. (1991). *EMBO J.* **10**, 3447-3456.

Kanaar, R., Lee, A. L., Rudner, D. Z., Wemmer, D. E. and Rio, D. C. (1995). *EMBO J.* **14**, 4530-4539.

Kang, C.-H., Chan, R., Berger, I., Lockshin, C., Green, L., Gold, L. and Rich, A. (1995). *Science* **268**, 1170-1173.

Kay, L. E., Muhandiram, D. R., Farrow, N. A., Aubin, Y. and Forman-Kay, J. D. (1996). *Biochemistry* **35**, 361-368.

Kenan, D. J., Query, C. C. and Keene, J. D. (1991). *TIBS* **16**, 214-220.

Kharrat, A., Macias, M. J., Gibson, T. J., Nilges, M. and Pastore, A. (1995). *EMBO J.* **14**, 3572-3584.

Kim, J. L., Nikolov, D. B. and Burley, S. K. (1993). *Nature* **365**, 520-527.

Kim, Y., Geiger, J. H., Hahn, S. and Sigler, P. B. (1993). *Nature* **365**, 512-520.

Laird-Offringa, I. A. and Belasco, J. G. (1995). *Proc. Natl. Acad. Sci. USA* **92**, 11859-11863.

LeCuyer, K. A., Behlen, L. S. and Uhlenbeck, O. C. (1996). *EMBO J.* **15**, 6847-6853.

Liljas, A. and Garber, M. (1995). *Curr. Op. Struct. Biol.* **5**, 721-727.

Love, J. J., Li, X., Case, D. A., Giese, K., Grosschedl, R. and Wright, P. E. (1995). *Nature* **376**, 791-795.

Markus, M. A., Hinck, A. P., Huang, S., Draper, D. E. and Torchia, D. A. (1997). *Nature Struc. Biol.*, **4**, 70-77.

Martinez, A., Burns, C. M. and Richardson, J. P. (1996a) *J. Mol. Biol.* **257**, 909-918.

Martinez, A., Opperman, T. and Richardson, J. P. (1996b) *J. Mol. Biol.* **257**, 895-908.

Mohr, G., Zhang, A., Gianelos, J. A., Belfort, M. and Lambowitz, A. M. (1992). *Cell* **69**, 483-494.

Mohr, G., Caprara, M. G., Guo, Q. and Lambowitz, A. M. (1994). *Nature* **370**, 147-150.

Musco, G., Stier, G., Joseph, C., Castiglione Morelli, M. A., Nilges, M., Gibson, T. J. and Pastore, A. (1996). *Cell* **85**, 237-245.

Nagai, K. (1996). *Curr. Op. Struct. Biol.* **6**, 53-61.

Nagai, K., Oubridge, C., Jessen, T. H., Li, J. and Evans, P. R. (1990). *Nature* **348**, 515-520.

Nagai, K., Oubridge, C., Ito, N., Avis, J. and Evans, P. (1995). *TIBS* **20**, 235-240.

Oubridge, C., Ito, N., Evans, P. R., Teo, C.-H. and Nagai, K. (1994). *Nature* **372**, 432-438.

Policarpou-Schwartz, M., Gunderson, S. I., Kandels-Lewis, S., Séraphin, B. and Mattaj, I. W. (1996). *RNA* **2**, 11-23.

Puglisi, J. D., Chen, L., Blanchard, S. and Frankel, A. D. (1995). *Science* **270**, 1200-1203.

Query, C. C., Bentley, R. C. and Keene, J. D. (1989). *Cell* **57**, 89-101.

Rould, M. A., Perona, J. J., Söll, D. and Steitz, T. A. (1989). *Science* **246**, 1135-1142.

Rould, M. A., Perona, J. J. and Steitz, T. A. (1991). *Nature* **352**, 213-218.

Scherly, D., Boelens, W., Dathan, N. A., van Venrooij, W. J. and Mattaj, I. (1990). *Nature* **345**, 502-506.

Scherly, D., Kambach, C., Boelens, W., van Venrooij, W. J. and Mattaj, I. W. (1991). *J. Mol. Biol.* **219**, 577-584.

Schindelin, H., Marahiel, M. A. and Heinemann, U. (1993). *Nature* **364**, 164-168.

Schnuchel, A., Wiltscheck, R., Czisch, M., Herrier, M., Willimsky, G., Graumann, P., Marahiel, M. A. and Holak, T. A. (1993). *Nature* **364**, 169-171.

Seeman, N. C., Rosenberg, J. M. and Rich, A. (1976). *Proc. Natl. Acad. Sci. USA* **73**, 804-808.

Shamoo, Y., Abdul-Manam, N., Patten, A. M., Crawford, J. K., Pellegrini, M. C. and Williams, K. R. (1994). *Biochemistry* **33**, 8272-8281.

Siomi, H., Siomi, M. C., Nussbaum, R. L. and Dreyfuss, G. (1993). *Cell* **74**, 291-298.

Slijper, M., Boelens, R., Davis, A. L., Konings, R. N. H., van der Marel, G. A., van Boom, J. H. and Kaptein, R. (1997). *Biochemistry* **36**, 249-254.

Spolar, R. S. and Record, M. T. J. (1994). *Science* **263**, 777-784.

St Johnston, D., Brown, N. H., Gall, J. G. and Jantsch, M. (1992). *Proc. Natl. Acad. Sci. USA* **89**, 10979-10983.

Steitz, T. A. (1990). *Q. Rev. Biophys.* **23**, 205-280.

Tacke, R. and Manley, J. L. (1995). *EMBO J.* **14**, 3540-3551.

Tsai, D. E., Harper, D. S. and Keene, J. D. (1991). *Nucleic Acids Res.* **19**, 4931-4936.

Valegård, K., Murray, J. B., Stockley, P. G., Stonehouse, N. J. and Liljas, L. (1994). *Nature* **371**, 623-626.

van Gelder, C. W. G., Gunderson, S. I., Jansen, E. J. R., Boelens, W. C., Polycarpou-Schwartz, M., Mattaj, I. W. and van Venrooij, W. J. (1993). *EMBO J.* **12**, 5191-5200.

Weeks, K. M. and Cech, T. R. (1995). *Cell* **82**, 221-230.

Weeks, K. M. and Cech, T. R. (1996). *Science* **271**, 345-348.

Werner, M. H., Huth, J. R., Gronenborn, A. M. and Clore, G. M. (1995). *Cell* **81**, 705-714.

Xing, Y., GuhaThakurta, D. and Draper, D. E. (1997). *Nature Struc. Biol.* **4**, 24-27.

Ye, X., Kumar, R. A. and Patel, D. J. (1995). *Chemistry & Biology* **2**, 827-840.

Ye, X., Gorin, A., Ellington, A. D. and Patel, D. J. (1996). *Nature Struc. Biol.* **3**, 1026-1033.

III

RNA AS A TARGET/ RNA AS A TOOL

6

Exploiting RNA: Potential Targets, Screens and Drug Development

CATHERINE D. PRESCOTT, LISA HEGG, KELVIN NURSE, RICHARD GONTAREK, HU LI, VICTORIA EMERICK, THERESE STERNER, MICHAEL GRESS, GEORGE THOM, SABINE GUTH AND DONNA RISPOLI

SmithKline Beecham Pharmaceuticals, 709 Swedeland Road, King of Prussia, PA 19406, USA

Ribosomes contain both protein and nucleic acid, one is the actor the other the supporting cast; RNA is moving into the spotlight. (Holtzman, 1985)

Abstract

The widespread emergence of acquired resistance to antibiotics in bacteria constitutes a serious problem to public health. The rapidity with which bacteria evolve drug resistance has further contributed to the drive for new research strategies to discover and develop novel chemotherapeutics. One approach for the development of new anti-infectives is the systematic exploitation of RNA as a novel target class. The functional importance of RNA is by now well established: RNA participates in almost all macromolecular processes including RNA processing, protein synthesis and protein transport. Natural inhibitors of protein synthesis are well documented and therefore ribonucleoprotein particles (RNPs) can be considered as valid targets for drug development (Spahn and Prescott, 1996). Detailed knowledge of the structure and function of any target is key to understanding the mechanism of drug action. However, this is clearly difficult when faced with the size and complexity of RNPs involved in many of these processes. Several

THE MANY FACES OF RNA
ISBN 0-12-233210-5

multi-disciplinary approaches that address these problems and thereby enable the study of RNA–ligand interactions have been described (Schroeder, 1994; Purohit and Stern, 1994; Fourmy *et al.*, 1996). These are reviewed here in the context of their potential application to target selection and identification, the establishment of screens to search for new drugs and drug development required to address parameters such as selectivity, potency and toxicity.

1 Introduction

The discovery of antibiotics is arguably the single most important advance in the history of medicine. However, the extensive use and abuse of antibiotics has had devastating consequences such that the population is now faced with the problems associated with the rapid spread of drug-resistant pathogens. Furthermore, antibiotics available today are merely structural derivatives of long-discovered antibiotic classes. The penems, for example, one of the last major classes of novel antibiotics to be discovered, have been known for nearly 20 years. It is both the emergence of acquired resistance and the lack of truly novel antibiotics that have fuelled the need for the identification of new targets and strategic approaches to drug discovery. One such approach is to systematically exploit RNA, which until recently has received only limited attention within the pharmaceutical industry. The aim of this article is to examine why RNA is considered to be a suitable target, what issues need to be addressed so as to exploit this class of molecules, and to review some of the approaches as to how RNA is being exploited both as a target and tool to aid drug discovery.

2 Validity of RNA as a target

The functional importance of RNA is by now well established, participating in almost all macromolecular processes. RNA motifs regulate RNA processing, mRNA translation, stability and localization. Numerous examples of catalytic RNAs have been cited including RNase P, the small catalytic RNAs (for example, the hammerhead ribozyme) and group I introns. The most extensively studied RNAs are conceivably those that constitute the ribosome: the rRNAs are critical to both the structure and function of the ribosome during protein synthesis. These are all examples of potential RNA-based targets. The validity of RNA as a target is exemplified by the evolution of aminoglycosides that interact directly with RNA and the effect of point mutations within the rRNAs that confer drug resistance (Davies *et al.*, 1993; Schroeder and von Ahsen, 1996). Small molecule inhibitors of group I intron self-splicing include guanosine analogues (Bass and Cech, 1984; Bass and Cech, 1986), arginine (reviewed by Yarus, 1993) and certain aminoglycosides (Lin *et al.*, 1992; Liu *et al.*, 1994). Neomycin has been shown to inhibit the hammerhead (Clouet d'Orval *et al.*, 1995; Stage *et al.*, 1995) and hepatitis delta virus ribozymes (Rogers *et al.*, 1996) and the tuberactinomycin family of antibiotics inhibits group I intron RNA splicing (Wank *et al.*, 1994). Although RNA is a charged molecule which could be predicted to interact non-specifically with antibiotics, *in vitro* selection experiments indicate that RNA–aminoglycoside interactions are not solely ionic (Wallis *et al.*, 1995; Schroeder and von Ahsen, 1996). These data, coupled with the observation that not all catalytic RNAs and functional ribonucleoproteins (RNPs) are inhibited by aminoglycosides (Schroeder and von Ahsen, 1996), suggest that there is a great deal to learn about small molecule binders and specific inhibitors of functional and catalytic RNA.

3 Target selection: criteria and issues

Selection of a target for exploitation is based on a number of criteria including essentiality, selectivity and spectrum. This requires confirmation of the functional importance of the target to the pathogen and assurance that a structural homologue does not exist in the infected host since this could also be inhibited by the antibiotic. A multidisciplinary approach is required to address these questions including molecular biology, biochemistry and more recently, bioinformatics. Significant progress has been made in sequencing genomes of several organisms, including those of the bacteria *E. coli, B. subtilis* and *H. influenzae*. Sequence similarity searching is a major resource that enables function to be inferred by homology to other sequences in the databases. However, a limitation of such analysis is reflected by those genes for which a match defines the biochemical function but the physiological function remains obscure. The potential of the sequencing projects may only be realized by a parallel experimental approach to functional analysis. The ability to characterize a functional aspect of the target is key to the development of a screen with which to identify inhibitory compounds. For example, Mei *et al.* (1996) have devised a high-throughput screen based on the functional activity of the self-splicing group I intron derived from *Pneumocystis carinii*. The desirable properties of these compounds (for example potency, penetration and stability) can potentially be improved by rational design, if the target is amenable to detailed structural investigation. While these parameters are true for the selection of any target, additional factors unique to RNA and its interactions must be considered.

RNA plays a number of functional roles and it is apparent that these are mediated by signals that are a combination of both sequence and structure motifs. The structural versatility of RNA is enormous, forming for example, helices, hairpins, pseudoknots, internal bulges and loop regions. The major groove of the RNA A form helix is deep and narrow, and therefore relatively inaccessible to ligand binding. Disruption of the helix by a bulged nucleotide or internal loop results in a widened groove and consequently allows access for a ligand to bind. Therefore an understanding of both RNA sequence and structure is essential. This has an impact on data retrieval from the genomic sequence databases: sequence similarity searching must include the ability to search for structural motifs which is dependent on prior identification of these motifs. Programs that search specifically for RNA families (for example, tRNAs and catalytic introns) and RNA structures (for example, regulatory motifs) are available (reviewed by Dandekar and Hentze, 1995) and have been successful in identifying new members of several RNA families. Accordingly, it is now feasible to apply computational approaches to identify novel RNA motifs.

The intrinsic structural flexibility of RNA makes biophysical approaches only possible for unusually conformationally stable RNA molecules. RNAs which are conformationally restrained by proteins tend to be within large macromolecular complexes and therefore also not amenable to biophysical analysis. A variety of approaches have been developed to address these issues and thereby permit detailed investigation of the structure and function of RNA molecules and which are in turn applicable to the drug discovery process. The following sections describe a number of these technologies and their application to anti-infective targets.

4 Exploiting RNA as a target and tool

4.1 Fragmentation

The size and complexity of RNA and ribonucleoprotein particles has hindered their detailed structural and functional analyses. Therefore, their fragmentation into functional subdomains that are amenable to detailed investigation is a potent strategy to advance RNA-based research (Schroeder, 1994). The ribosome, for example, is naturally organized into subdomains: specific RNP particles encompassing the 5′, 3′ and central domains of the 30 S subunit have been isolated (Zimmerman, 1974; Yuki and Brimacombe, 1975) and also reassembled *in vitro* (Nomura *et al.*, 1969; Weitzmann *et al.*, 1993; Samaha *et al.*, 1994). Further dissection of the ribosome has been achieved and expanded to include interactions with additional ligands: an oligoribonucleotide analogue of the decoding region located near to the 3′ end of 16 S rRNA has been shown to interact with both antibiotic (neomycin) and RNA ligands (tRNA and mRNA) of the 30 S subunit in a manner that resembles normal subunit function (Purohit and Stern, 1994).

While the above-mentioned approaches have been studied *in vitro*, an *in vivo* approach has also been developed to study RNA–ligand interactions. The 'RNA fragment rescue' assay is based on the premise that the *in vivo* expression of RNA fragments able to bind and thereby sequester a drug permits the continued functioning of the intact enzyme and thus ensures cell viability. This was demonstrated by the expression of a 16 S rRNA fragment encompassing the spectinomycin binding domain that conferred drug resistance. The potential of this approach was further emphasized by the identification of additional sequences from among a mixed population that conferred spectinomycin resistance (Howard *et al.*, 1995; Thom and Prescott, 1997). The sequences were predicted to share common structural motifs required for drug binding. The ability to screen RNA molecules in this manner and subsequently identify the minimal structural motifs involved in intermolecular interactions is clearly an advantage to drug development.

4.2 SELEX technologies

SELEX, the Systematic Evolution of Ligands by EXponential enrichment, is an extremely powerful *in vitro* selection procedure that can lead to the identification of minimal structural motifs that characterize, for example, a ligand binding site (Tuerk and Gold, 1990). This approach has been used successfully to characterize the interaction of naturally occurring nucleic acids with proteins, to generate novel nucleic acid binding species called aptamers that in some instances act as antagonists (Gold *et al.*, 1995), and to investigate the catalytic power of RNA (Lehman and Joyce, 1993; Lorsch and Szostak, 1994; Illangasekare *et al.*, 1995; Wilson and Szostak, 1995). A variety of molecules have been targeted including nucleic acid binding proteins (Tuerk and Gold, 1990; Bartel *et al.*, 1991; Schneider *et al.*, 1992; Tuerk *et al.*, 1992), non-nucleic acid binding proteins (Bock *et al.*, 1992; Jellinek *et al.*, 1993; Nieuwlandt *et al.*, 1995; Binkley *et al.*, 1995) and small molecules (Ellington and Szostak, 1990; Famulok and Szostak, 1992; Connell *et al.*, 1994; Jenison *et al.*, 1994; Louhon and Szostak, 1995).

RNA molecules that bind with high affinity and specificity have been exploited as tools. Aptamers that behave as antagonists have been selected to proteases, growth factors (Gold *et al.*, 1995) and antiviral targets. Several laboratories have carried out *in vitro* selection using the HIV-1 Rev protein as the target (Bartel *et al.*, 1991; Giver *et al.*, 1993; Tuerk and MacDougal-Waugh, 1993; Jensen *et al.*, 1995); the evolved RNA ligands from these studies reveal striking sequence and secondary structural similarities to the previously identified high-affinity Rev-binding site on the Rev-responsive element (RRE) RNA Stem IIB (Heaphy *et al.*, 1990; Heaphy *et al.*, 1991). More recently, Jensen *et al.* (1995) have described an *in vitro* selection procedure called crosslinking SELEX, where a randomized RNA library substituted with the photoreactive chromophore 5-iodouracil was irradiated by UV light in the presence of Rev. This method of selection resulted in RNA molecules which exhibited subnanomolar dissociation constants and high-efficiency photocrosslinking to Rev. Interestingly, some of the aptamers isolated by this procedure formed a stable complex with Rev that was resistant to denaturing gel electrophoresis in the absence of UV irradiation. This was taken to indicate the possibility of covalent attachment of the aptamer to the protein and further suggests the potential to identify nucleic acid molecules with a capacity to covalently attack a target molecule.

Another novel application of *in vitro* selection–amplification procedures was described by Pan *et al.* (1995). In this report, rather than using a purified protein as a target, intact retrovirus particles (Rous sarcoma virus, RSV) were used to isolate RNA and nuclease-resistant RNA analogues that specifically bound to and neutralized the virus. Using intact viral particles as targets may be advantageous because it circumvents the need for a full understanding of the complex mechanisms of viral infection, and may provide selection against a target protein in its native, complexed state rather than as a single, purified entity. These novel *in vitro* selection–amplification techniques represent a promising approach towards identifying small nucleic acid ligands that may act as specific viral inhibitors.

SELEX-derived oligonucleotides are also suitable as diagnostic tools. Davis *et al.* (1997) demonstrated that fluoresceinated DNA aptamers were equally effective as an anti-human neutrophil elastase antibody. Accordingly, SELEX-derived oligonucleotides may be developed as universal diagnostic reagents. Although antibodies are excellent diagnostic reagents, they are difficult to modify due to their size limitation (150 kD) and can not be raised against toxic or non-immunogenic targets. Oligonucleotides on the other hand, are small and have no target limitation. In addition, a variety of chemical modifications (for example fluorescein, biotin, radioisotopes) can be readily introduced into the oligonucleotide for detection purposes.

4.3 Antisense and ribozymes

Antisense tactics have been used to target RNA molecules and thereby block gene expression (Cohen, 1991). Several groups have engineered retroviral and non-retroviral vectors to express antisense RNAs against important HIV-1 RNAs (Chuah, *et al.*, 1994; Lori *et al.*, 1994; Cagnon *et al.*, 1995; Biasolo *et al.*, 1996), and have demonstrated that these constructs can indeed inhibit viral gene expression and replication. Similarly, it has been shown that specific domains within the HCV 5' non-coding region that are critical for translation are susceptible to antisense inhibition (Tsukiyama-Kohara *et al.*, 1992; Wang *et al.*, 1993; Wakita and Wands, 1994). Several laboratories have also expressed sense or 'decoy' RNAs believed to inhibit HIV-1 gene expression by sequestering viral RNA-binding regulatory proteins Tat and Rev (Lisziewicz *et al.*, 1993; Chang *et al.*, 1994; Lisziewicz *et al.*, 1995a,b; Bahner *et al.*, 1996; Kim *et al.*, 1996). Others have combined antisense gene therapy with retroviral vectors that express a mutant, transdominant Rev protein to interfere with the function of the virally expressed protein (Vandendriessche *et al.*, 1995; Morgan and Walker, 1996). Chemical modification of the antisense analogues can increase nuclease resistance resulting in effective inhibitors of HIV-1 expression (Anazodo *et al.*, 1995a,b; Weichold *et al.*, 1995).

Exploiting the catalytic properties of ribozymes for gene inactivation has several advantages over traditional antisense technologies. Delivery of catalytic amounts of ribozymes rather than molar excesses required by many antisense approaches is a distinct benefit. Because catalytic RNAs interact with the target by Watson-Crick base pairing, 'designer' ribozymes can be synthesized for targeted inhibition of gene function (Grassi and Marini, 1996). Additionally, as ribozymes typically break the phosphodiester backbone of the target molecule, inactivation is likely to make the target RNA more susceptible to cellular nucleases. Several studies have already demonstrated the feasibility of using ribozymes to reduce or eliminate RNAs of HCV (Lieber *et al*., 1996), HIV-1 (Yamada *et al*., 1996), and influenza A virus (Tang *et al*., 1994). Ribozyme delivery (exogenous or endogenous expression), stability and cellular localization are important issues that are subject to intense investigation. Several groups have increased ribozyme stability by the introduction of chemical modifications (Usman and Stinchcomb, 1996 and references within; Pagratis *et al*., 1997) or by circularization of the ribozyme (Puttaraju and Been, 1996). Directed gene inactivation by RNase P has also been reported. This involves the introduction of an RNA molecule (external guide sequence) that forms a tRNA-like structure with the RNA target which is consequently recognized and cleaved by the host RNase P (Forster and Altman, 1990; Surratt *et al*., 1990; Yuan *et al*., 1992). There is also a report of repair of a defective mRNA by a trans-splicing *Tetrahymena* group I intron (Sullenger and Cech, 1994).

4.4 RNA–ligand interactions

The importance of ribonucleoprotein complexes is well established. However, relatively little information is available that describes RNA–protein interactions at the molecular level. Several strategies have been reported that aid the identification and characterization of the key recognition elements involved in protein–RNA complex formation.

A translation-repressor assay was developed to investigate the binding of HIV Rev protein to the HIV RRE (Rev responsive element RNA) (Jain and Belasco, 1996). The HIV RRE was inserted immediately upstream of the reporter (β-galactosidase) Shine-Dalgarno sequence. Specific binding of the RRE by Rev protein sterically hindered the ability of the ribosome to bind to the Shine-Dalgarno sequence, resulting in decreased levels of the reporter. This system identified mutations in the RRE that altered Rev binding, as well as suppressor mutations in Rev that rescue the binding of altered RRE sequences. The results from this genetic screen supported previous findings that recognition of the RRE by Rev is mediated by the arginine rich α-helical domain of the protein, with sequence-specific interactions arising from amino acids on the non-arginine face of the helix. It is conceivable that this translation-repressor assay is applicable to the investigation of other RNA–protein complexes.

To further characterize the precise domain within a protein engaged in RNP complex formation, a novel screen has been reported based on the bacteriophage λ antitermination event (Franklin, 1993; Harada *et al*., 1996). Antitermination is initiated by the interaction between the arginine rich N-terminal domain of the N-protein and the nut RNA stem loop (Franklin, 1993; Tan and Frankel, 1995). Formation of this RNP complex acts as a signal for the recruitment of additional factors which together form an antitermination complex and consequently permit onward transcription. The system was modified to investigate the interaction of HIV Rev with RRE. Twenty-one amino acids of the N terminus of N protein were replaced with an arginine-biased peptide library and RRE inserted in place of the nut RNA stem loop (Harada *et al.*, 1996). Specific interaction of the HIV RRE with modified N protein resulted in expression of a β-galactosidase reporter gene, located downstream of terminator sequences. Analysis of the modified N proteins able to elicit antitermination showed that the N-terminal region contained sequences matching the Rev consensus sequence. This system is a powerful method to rapidly screen and identify critical residues involved in protein–RNA interactions.

Phage display may be another powerful approach to identify key residues involved in binding to RNA. Phage display involves the expression of an enormous variety of peptides or proteins on the surface of filamentous phage (Smith, 1985; Cwirla *et al.*, 1990; Devlin *et al.*, 1990; Scott and Smith, 1990) which can be efficiently screened against a target without testing each member individually (Sternberg and Hoess, 1995). Typically this has been deployed to study protein–protein interactions (reviewed by Burton, 1995). However Belasco and colleagues (Laird-Offringa and Belasco, 1995) have used phage display technologies to create a library of peptides based on the U1A protein which was screened against U1 hairpin II target RNA. A key amino acid required for complex formation was identified along with a sequence motif characteristic of the U1A protein. Although this is the first report of phage display screened against RNA, it demonstrates the feasibility of this exciting approach to further investigate RNA–ligand interactions.

4.5   Biophysical approaches

A wealth of information regarding the secondary structure of RNA has been accumulated from mutational analysis and chemical and enzymatic probing studies and will not be reviewed here. Detailed secondary and tertiary structural information obtained by biophysical techniques has been made possible using structural analogues that circumvent the problems associated with the size and fluidity of RNA. The current literature is replete with data regarding both the structure of RNA molecules and ligand interactions. Therefore the following section focuses on the experimental achievements with respect to those RNAs of particular interest to the field of drug development.

Limitations associated with the intrinsic structural flexibility of RNA have been circumvented using both RNA–ligand complexes to conformationally restrict the molecule as well as structural analogues in which functional domains are commonly flanked by stabilizing elements. For example, data have been reported for RNP complexes between RNA and MS2 bacteriophage coat protein, tRNA synthetases (Valegard *et al.*, 1994), as well as HIV Tat–TAR and Rev–RRE complexes. Recent insights into the function of the ribosome's decoding region is an example of the advantage of using smaller structural analogues. The decoding site analogue has been investigated by NMR spectroscopy, yielding insights not only into the structure of the RNA within this region, but also into the details of aminoglycoside binding (Fourmy *et al*., 1996; Recht *et al*., 1996).

X-ray crystallography has proven to be an essential tool for providing a detailed understanding of molecular organization, knowledge paramount to the ability to determine any structure–function relationship. Crystallographic data is available, for example, on the hammerhead ribozyme and the catalytic domain of the group 1 ribozyme (reviewed in Moras, 1997) as well as fragments of 5 S rRNA (Leontis and Moore, 1986; Betzel *et al*., 1995). Other spectroscopic techniques such as NMR and circular dichroism have been extensively applied to study both structural and functional aspects of RNAs such as HIV RRE and TAR as well as the interactions with their counterpart proteins, Rev and Tat (Aboul-ela *et al*., 1995; Puglisi *et al*., 1995; Williamson *et al*., 1995; Metzger *et al*., 1996 and references therein).

The information gained by such biophysical analysis provides the potential for rational drug design as well as the development of screening methods to identify ligands to either disrupt or alter RNA-associated functions that play a key role in the propagation of infectious agents.

5 Concluding remarks

RNA plays a central role in controlling the genetic diversity within a cell, and is therefore an attractive anti-infective target. The systematic exploitation of RNA for drug development has been dependent on surmounting the problems associated with investigating large, complex and intrinsically flexible molecules. Innovative approaches have begun to meet these challenges, resulting in RNA research as being one of the most rapidly developing areas of science.

References

Aboul-ela, F., Karn, J. and Varani, G. (1995). *J. Mol. Biol.* **253**, 313–332.

Anazodo, M. I., Salomon, H., Friesen, A. D., Wainberg, M. A. and Wright, J. A. (1995a). *Gene* **166**, 227–232.

Anazodo, M. I., Wainberg, M. A., Friesen, A. D. and Wright, J. A. (1995b). *J. Virol.* **69**, 1794–1801.

Bahner, I., Kearns, K., Hao, Q. L., Smogorzewska, E. M. and Kohn, D. B. (1996). *J. Virol.* **70**, 4352–4360.

Bartel, D. P., Zapp, M. L., Green, M. R. and Szostak, J. W. (1991). *Cell* **67**, 529–536.

Bass, B. L. and Cech, T. R. (1984). *Nature* **308**, 820–826.

Bass, B. L. and Cech, T. R. (1986). *Biochemistry* **25**, 4473–4477.

Betzel, C., Lorenz, S., Furste, J. P., Bald, R., Zhang, M., Schneider, T., Wilson, K. and Erdmann, V. A. (1995). *FEBS Lett.* **351**, 159–164.

Biasolo, M. A., Radaelli, A., Del Pup, L., Franchin, E., De Giuli-Morghen, C. and Palu, G. (1996). *J. Virol.* **70**, 2154–2161.

Binkley, J., Allen, P., Brown, D. M., Green, L., Tuerk, C. and Gold, L. (1995). *Nucleic Acids. Res.* **23**, 3198–3205.

Bock, L., Griffin, L. Latham, J., Vermass, E. and Toole, J. (1992). *Nature* **355**, 546–566.

Burton, D. R. (1995). *Immunotech.* **1**, 87–94.

Cagnon, L., Cucchiarini, M., Lefebvre, J. C. and Doglio, A. (1995). *J. Acquir. Immune Defic. Syndr. Hum. Retrovirol.* **9**, 349–358.

Chang, H. K., Gendelman, R., Lisziewicz, J., Gallo, R. C. and Ensoli, B. (1994). *Gene Ther.* **1**, 208–216.

Chuah, M. K., Vandendriessche, T., Chang, H. K. and Morgan, R. A. (1994). *Hum. Gene Ther.* **5**, 1467–1475.

Clouet d'Orval, B., Stage, T. K. and Uhlenbeck, O. C. (1995). *Biochemistry* **34**, 11186–11190.

Cohen, J. S. (1991). *Antiviral Res.* **16**, 121–133.

Connell, G. J., Illangeskare, M. and Yarus, M. (1994). *Science* **264**, 1137–1141.

Cwirla, S. E., Peters, E. A., Barrett, R. W. and Dower, W. J. (1990). *Proc. Natl. Acad. Sci. USA* **87**, 6378–6382.

Dandekar, T. and Hentze, M. W. (1995). *TIG* **11**, 45–50.

Davies, J., von Ahsen, U. and Schroeder, R. (1993). In *The RNA World* (R. F. Gesteland and J. F. Atkins, eds), pp. 185–204. Cold Spring Harbor Laboratory Press, Cold Spring Harbor, NY.

Davis, K. A., Abrams, B., Lin, Y. and Jayasena, S. D. (1997). *Nucleic Acids Res.* **24**, 702–706.

Devlin, J. J., Panganiban, L. C. and Devlin, P. E. (1990). *Science* **249**, 404–406.

Ellington, A. D. and Szostak, J. W. (1990). *Nature* **346**, 818–822.

Famulok, M. and Szostak, J. W. (1992). *J. Am. Chem. Soc.* **114**, 3990–3991.

Forster, A. C. and Altman, S. (1990). *Science* **249**, 783–785.

Fourmy, D., Recht, M. I., Blanchard, S. C. and Puglisi, J. D. (1996). *Science* **274**, 1367–1371.

Franklin, N. C. (1993). *J. Mol. Biol.* **231**, 343–360.

Giebel, L. B., Cass, R. T., Milligan, D. L., Young, D. C., Arze, R. and Johnson, C. R. (1995). *Biochemistry* **34**, 15430–15435.

Giver, L., Bartel, D., Zapp, M., Pawul, A., Green, M. and Ellington, A. D. (1993). *Nucleic Acids Res.* **21**, 5509–5516.

Gold, L. (1995). *J. Biol. Chem.* **270**, 13581–13584.

Gold, L., Polisky, B., Uhlenbeck, O. and Yarus, M. (1995). *Annu. Rev. Biochem.* **64**, 763–797.

Gold, L., Brown, D., He, Y., Shtatland, T., Singer, B. and Wu, Y. (1997). *Proc. Natl. Acad. Sci. USA* **94**, 59–64.

Grassi, G. and Marini, J. C. (1996). *Annals of Medicine* **28**, 499–510.

Harada, K., Martin, S. S. and Frankel, A. D. (1996). *Nature* **380**, 175–179.

Heaphy, S., Dingwall, C., Ernberg, I., Gait, M. and Skinner, M. A. (1990). *Cell* **60**, 685–693.

Heaphy, S., Finch, J. T., Gait, M., Karn, J. and Singh, M. (1991). *Proc. Natl. Acad. Sci. USA* **88**, 7366–7370.

Holzman, D. (1985). *Mosaic* **16**, 33–38.

Howard, B., Thom, G., Jeffrey, I., Colthurst, D., Knowles, D. and Prescott, C. D. (1995). *Biochem. Cell Biol.* **73**, 1161–1166.

Illangasekare, M., Sanchez, G., Nickeles, T. and Yarus, M. (1995). *Science* **267**, 643–647.

Jain, C. and Belasco, J. G. (1996). *Cell* **87**, 115–125.

Jellinek, D., Lynott, C. K., Rifkin, D. B. and Janjic, N. (1993). *Proc. Natl. Acad. Sci. USA* **90**, 11227–11227.

Jenison, R. D., Gill, S. C., Pardi, A. and Polisky, B. (1994). *Science* **263**, 1425–1429.

Jensen, K. B., Atkinson, B. L., Willis, M. C., Koch, T. H. and Gold, L. (1995). *Proc. Natl. Acad. Sci. USA* **92**, 12220–12224.

Katz, B. A. (1995). *Biochemistry* **34**, 15421–15429.

Kim, J. H., McLinden, R. J., Mosca, J. D., Vahey, M. T., Greene, W. C. and Redfield, R. R. (1996). *J. Acquir. Immune Defic. Syndr. Huqm. Retrovirol.* **12**, 343–351.

Laird-Offringa, I. A and Belasco, J, G. (1995) *Proc. Natl. Acad. Sci. USA* **92**, 11859–11863.
Laird-Offringa, I. A. and Belasco, J. (1996). *Methods in Enzymology* **267**, 149–168.
Lehman, N. and Joyce, G. F. (1993). *Nature* **361**, 182–185.
Leontis, N. B. and Moore, P. B. (1986). *Biochemistry* **25**, 3916–3925.
Lieber, A., Cheng-Yi, H., Polyak, S. J., Gretch, D. R., Barr, D. and Kay, M. A. (1996). *J. Virol.* **70**, 8782–8791.
Lin, H., Niu, M. T., Yoganathan, T. and Buck, G. A. (1992). *Gene* **119**, 163–173.
Lisziewicz, J., Sun, D., Smythe, J., Lusso, P., Lori, F., Louie, A., Markham, P., Rossi, J., Reitz, M. and Gallo, R. C. (1993). *Proc. Natl. Acad. Sci. USA* **90**, 8000–8004.
Lisziewicz, J., Sun, D., Lisziewicz, A. and Gallo, R. C. (1995a). *Gene Ther.* **2**, 218–222.
Lisziewicz, J., Sun, D., Trapnell, B., Thomson, M., Chang, H. K., Ensoli, B. and Peng, B. (1995b). *J. Virol.* **69**, 206–212.
Liu, Y., Tidwell, R. R. and Leibowitz, M. J. (1994). *J. Euk. Microbiol.* **41**, 31–38.
Lori, F., Lisziewicz, J., Smythe, J., Cara, A., Bunnag, T. A., Curiel, D. and Gallo, R. C. (1994). *Gene Ther.* **1**, 27–31.
Lorsch, J. R. and Szostak, J. W. (1994). *Nature* **371**, 31–36.
Louhon, C. T and Szostak, J. W. (1995). *J. Am. Chem. Soc.* **117**, 1246–1257.
Mei, H-Y., Cui, M., Sutton, S. T., Troung, H. N., Chung, F-Z. and Czarnik, A. W. (1996). *Nucleic Acids Res.* **24**, 5051–5053.
Metzger, A. U., Schindler, T., Willbold, D., Kraft, M., Steeborn, C., Volkman, A., Frank, R. W. and Rosch, P. (1996). *FEBS Lett.* **384**, 255–259.
Moras, D. (1997). *RNA* **3**, 111–113.
Morgan, R. A. and Walker, R. (1996) *Hum. Gene Ther.* **7**, 1281–1306.
Nieuwlandt, D., Wecker, M. and Gold, L. (1995). *Biochem.* **34**, 5651–5659.
Nomura, M., Traub, P., Buthrie, C. and Nashimoto, H. (1969). *J. Cell Physiol.* **74**, 241–252.
Pagratis, N., Bell, C., Chang, Y.-F., Jennings, S., Fitzwater, T., Jellinek, D. and Dang., C. (1997). *Nature Biotechnology* **15**, 68–73.
Pan, W., Craven, R. C., Qiu, Q., Wilson, C., Wills, J. W., Golovine, S. and Wang, J.-F. (1995). *Proc. Natl. Acad. Sci. USA* **92**, 11509–11513.
Puglisi, J. D., Tan, R., Calnan, B. J., Frankel, A. D. and Williamson, J. R. (1992). *Science* **257**, 76–80.
Puglisi, J. D., Chen, L., Blanchard, S. and Frankel, A. D. (1995). *Science* **270**, 1200–1203.
Purohit, P. and Stern, S. (1994). *Nature* **370**, 659–662.
Puttaraju, M. and Been, M. D. (1996). *SAAS Bull Biochem. Biotechnol.* **9**, 77–82.
Recht, M. I., Fourmy, D., Blanchard, S. C., Dahlquist, K. D. and Puglisi, J. D. (1996). *J. Mol. Biol.* **262**, 421–426.
Rogers, J., Chang, A. H., von Ahsen, U., Schroeder, R. and Davies, J. (1996). *J. Mol. Biol.* **259**, 916–925.
Samaha, R. R., O'Brien, T., O'Brien, B. and Noller, H. F. (1994). *Proc. Natl. Acad. Sci. USA* **91**, 7884–7888.
Schneider, D., Tuerk, C. and Gold, L. (1992). *J. Mol. Biol.* **228**, 862–869.
Schroeder, R. (1994). *Nature* **370**, 597–598.
Schroeder, R. and von Ahsen, U. (1996). In *Catalytic RNA: Nucleic Acids and Molecular Biology* **Vol 10** (F. Eckstein and D. M. J. Lilley, eds), pp. 53–74. Springer-Verlag, New York, NY.
Scott, J. K. and Smith, G. P. (1990). *Science* **249**, 386–390.
Smith, G. P. (1985). *Science* **228**, 1315–1317.
Spahn, C. M. T. and Prescott, C. D. (1996). *J. Mol. Med.* **74**, 423–439.
Stage, T. K., Hertel, K. J. and Uhlenbeck, O. C. (1995). *RNA* **1**, 95–101.
Sternberg, N. and Hoess, R. H. (1995). *Proc. Natl. Acad. Sci. USA* **92**, 1609–1613.
Sullenger, B. A. and Cech, T. R. (1994). *Nature* **371**, 619–622.

Surratt, C. K., Lesnikowski, Z., Schifman, A. L., Schmidt, F. J. and Hecht, S. M. (1990). *J. Biol. Chem.* **265**, 22506–22512.

Tan, R. and Frankel, A. D. (1995). *Proc. Natl. Acad. Sci. USA* **92**, 5282–5286.

Tang, X. B., Hobom, G. and Luo, D. (1994). *J. Med. Virol.* **42**, 385–395.

Thom, G. and Prescott, C. D. (1997). *Bioorg. Med. Chem.* in press.

Tsukiyama-Kohara, K., Iizuka, N., Kohara, M. and Nomoto, A. (1992). *J. Virol.* **66**, 1476–1483.

Tuerk, C. and Gold, L. (1990). *Science* **249**, 505–510.

Tuerk, C., MacDougal, S. and Gold, L. (1992). *Proc. Natl. Acad. Sci. USA* **89**, 6988–6992.

Tuerk, C. and MacDougal-Waugh, S. (1993). *Gene* **137**, 33–39.

Usman, N. and Stinchcomb, D. T. (1996). In *Catalytic RNA: Nucleic Acids and Molecular Biology* **Vol. 10** (F. Eckstein and D. M. J. Lilley, eds), pp. 243–281. Springer-Verlag, New York, NY.

Valegard, K., Murray, J., Stockley, P., Stonehouse, N. and Liljas, L. (1994). *Nature* **371**, 623–626.

Vandendriessche, T., Chuah, M. K., Chiang, L., Chang, H. K., Ensoli, B. and Morgan, R. A. (1995). *J. Virol.* **69**, 4045–4052.

Wakita, T. and Wands, J. R. (1994). *J. Biol. Chem.* **269**, 14205–14210.

Wallis, M. G., von Ahsen, U., Schroeder, R. and Famulok, M. (1995). *Chem. Biol.* **2**, 543–552.

Wang, C., Sarnow, P. and Siddiqui, A. (1993). *J. Virol.* **67**, 333–334.

Wank, H., Rogers, J., Davies, J. and Schroeder, R. (1994). *J. Mol. Biol.* **236**, 1001–1010.

Watanabe, G., Kawai, G., Muto, Y., Wantanabe, K., Inoue, T. and Yokoyama, S. (1996). *Nucleic Acids Res.* **24**, 1337–1344.

Weichold, F. F., Lisziewicz, J., Zeman, R. A., Nerurkar, L. S., Agrawal, S., Reitz, M. S. and Gallo, R. C. (1995). *AIDS Res. Hum. Retroviruses* **11**, 863–867.

Weitzmann, C. J., Cunningham, R. R., Nune, K. and Offengand, J. (1993). *FASEB J.* **7**, 177–180.

Williamson, J. R., Battiste, J. L., Mao, M. and Frankel, A. D. (1995). *Nucleic Acids Symp. Ser.* **33**, 46–48.

Wilson, C. and Szostak, J. W. (1995). *Nature* **374**, 777–782.

Yamada, O., Kraus, G., Luznik, L., Yu, M. and Wong-Staal, F. (1996). *J. Virol.* **70**, 1596–1601.

Yarus, M. (1993). In *The RNA World* (R. F. Gesteland and J. F. Atkins, eds), pp. 205–218. Cold Spring Harbor Laboratory Press, Cold Spring Harbor, NY.

Yuan, Y., Hwang, E-S. and Altman, S. (1992). *Proc. Natl. Acad. Sci. USA* **89**, 8006–8010.

Yuki, A. and Brimacombe, R. (1975). Eur. J. *Biochem.* **56**, 23–34.

Zimmerman, R. A. (1974). In *Ribosomes* (M. Nomura., A. Tissieres and P. Lengyel, eds), pp. 225–269. Cold Spring Harbor Laboratory Press, Cold Spring Harbor, N.Y., USA.

7

Structural Basis for Aminoglycoside Antibiotic Action*

JOSEPH D. PUGLISI

Center for Molecular Biology of RNA, University of California, Santa Cruz, CA 95064, USA

Abstract

Ribosomal RNA is the target for many antibiotics which interfere with the essential functions of protein synthesis. Aminoglycoside antibiotics bind directly to 16 S ribosomal RNA in the 30 S subunit and cause misreading of the genetic code. Biochemical experiments have localized the binding site of aminoglycosides to the region of codon–anticodon interaction in the ribosomal A-site. A 27 nt RNA that corresponds to the small subunit A site mimics the affinity and specificity of aminoglycoside binding to the ribosome. We have determined the solution structure of a complex between the aminoglycoside paromomycin and this RNA. Paromomycin binds in the major groove of the RNA within a pocket formed by a non-canonical A–A pair, and a bulged adenosine. Paromomycin makes specific contacts mediated by rings I and II, which are common to 11 aminoglycoside antibiotics that bind to this region of the ribosome. Rings III and IV, which vary and can be deleted while maintaining antibiotic activity, are not well defined in the solution structure and make non-specific RNA contacts. The aminoglycosides ribostomycin and neamine, which lack ring IV and rings III and IV, respectively, bind in a similar manner to the oligonucleotide. The structure explains the specificity of these antibiotics for prokaryotic ribosomes, as eukaryotic ribosomal RNAs are unable to form the specific binding pocket for aminoglycosides. Resistance to aminoglycosides is imparted by enzymatic modification of either the antibiotic or the RNA target, and the sites of these modifications are at the RNA–antibiotic interface. The structure represents a starting point for improved aminoglycoside antibiotics. Comparison of the structure of the free and bound

---

*Figure 6 for this chapter appears in the colour plate section between p. 144 and p. 145

THE MANY FACES OF RNA
ISBN 0-12-233210-5

RNA suggests a mechanism for aminoglycoside-induced miscoding, and experiments are in progress to test this mechanism.

1 Introduction

The ribosome is a ribonucleoprotein particle that catalyses protein synthesis. The ribosome has three major functions, which are decoding of the genetic code through the codon–anticodon interaction, catalysis of peptide bond formation, and translocation down the messenger RNA to continue the synthesis of the peptide chain. The accuracy and efficiency by which these tasks are completed is crucial to the life of a cell, which has evolved the complex structure of the ribosome to solve this problem. Unfortunately, there is as yet no high-resolution crystal structure of the ribosome to reveal its mysteries.

RNA is thought to be the essential component for ribosome function (Noller, 1991; Noller *et al*., 1992). 16 S and 23 S ribosomal RNAs, which are the components of the small and large ribosomal subunits, respectively, contain very highly conserved regions when ribosomes of different organisms are compared (Gutell, 1994). In contrast, the sequences and identities of ribosomal proteins vary strongly. The regions of strong conservation within ribosomal RNA often overlap with functional sites identified by biochemical experiments. Ribosomal proteins can be deleted without strong phenotype, whereas ribosomal RNA mutations and modifications can disrupt ribosome function.

Decoding of genetic information occurs on the small ribosomal subunit. Adjacent mRNA codons interact with the peptidyl tRNA and aminoacyl tRNA in the P and A sites, respectively. The nucleotides within 16 S ribosomal RNA that form these sites have been mapped by chemical probing methods (Moazed and Noller, 1990). The P-site is formed by diverse elements in 16 S ribosomal RNA, consistent with the high affinity of peptidyl tRNA for the P-site. In contrast, A-site bound tRNA protects a limited number of nucleotides in the so-called decoding site (Fig. 1a, b). Two crucial nucleotides, A1492 and A1493, are universally conserved in all ribosomal sequences. The high phylogenetic conservation of this region of 16 S ribosomal RNA is consistent with its important function in decoding. In addition, no ribosomal protein has been shown to interact directly with this region of ribosomal RNA (Powers and Noller, 1995).

Antibiotics target the ribosome and interfere with essential ribosome functions (Gale *et al*., 1981) such as initiation, decoding fidelity, peptidyl transfer, and translocation. Beyond their obvious pharmacological importance, antibiotics have been a powerful tool to dissect ribosome function. Many ribosome-directed antibiotics protect nucleotides within ribosomal RNA from reaction with chemical probes (Moazed and Noller, 1987a, b; Woodcock *et al*., 1991). In addition, antibiotic resistance often results from ribosomal RNA mutations and modification (see below). Coupled with the functional importance of ribosomal RNA, this strongly suggests that antibiotics that target the ribosome interact directly with RNA. Our work has focused on an important class of antibiotics, the aminoglycosides.

(a)

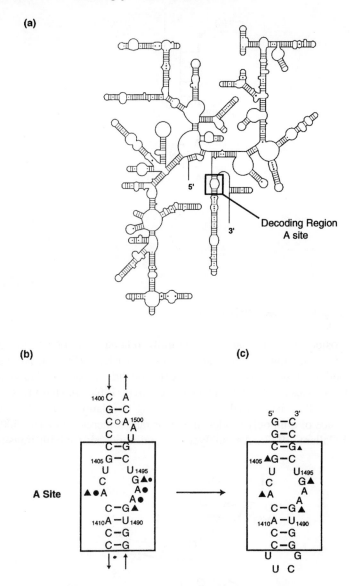

Fig. 1 (a) Secondary structure of *E. coli* 16 S ribosomal RNA. The decoding region A site, where the codon–anticodon interaction occurs is boxed. (b) Secondary structure and nucleotide sequence in the decoding region. Nucleotides protected from DMS modification by the addition of A-site tRNA and mRNA are indicated by closed circles. Nucleotides protected by addition of aminoglycoside antibiotics are indicated by triangles. (c) Model oligonucleotide that corresponds to the decoding region A-site. Nucleotides in the oligonucleotide that are protected from reaction with DMS in the presence of 1 μM paromomycin are indicated by triangles. (d) RNA sequence elements critical for specific aminoglycoside binding to the A-site RNA oligonucleotide. Required nucleotides are shown explicitly; N, any nucleotide; N-N, any Watson-Crick base pair. For position 1495, either a U or G gives high affinity binding, and each presents a hydrogen bond acceptor in the major groove.

(d)

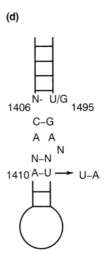

Fig. 1 (*Continued*)

2 Aminoglycoside antibiotics

Aminoglycoside antibiotics are a structurally related class of aminosugar and aminocyclitol-containing compounds that target bacteria (Fig. 2). Aminoglycoside antibiotics cause a marked decrease in the fidelity of translation (Davies *et al.*, 1965; Davies *et al.*, 1966; Davies and Davis, 1968). *In vivo* measurements have estimated that the normal error rate during translation is 1 in 10^4, whereas in the presence of aminoglycosides, the error rate is increased to 1 in 500 (Edelmann and Gallant, 1977). Aminoglycoside antibiotics also inhibit translocation.

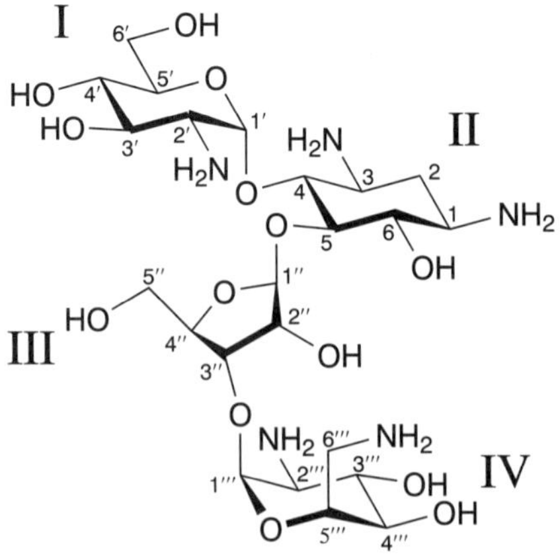

Fig. 2 Chemical structure of the aminoglycoside antibiotic paromomycin. Rings I and II are common elements among all aminoglycoside antibiotics.

The mechanism of the bactericidal action of aminoglycosides is not clear, but may be linked more towards the inhibition of translocation. Proteins are generally robust towards single mutations, whereas truncation caused by decreased ribosome processivity would be more deleterious (Kurland, 1992). Aminoglycosides bind to the conserved A-site region of 16 S ribosomal RNA. Chemical footprinting experiments, and RNA mutations and modifications that yield aminoglycoside resistance, confirm the direct binding of aminoglycosides to the 16 S ribosomal RNA in the small subunit.

Aminoglycoside antibiotics have common features for interaction with 16 S ribosomal RNA. All aminoglycosides that bind to the ribosomal A-site contain rings I and II; in particular, the two amino groups of ring II and a hydrogen bond donor at position 6′ in ring I are common elements. One or two additional rings can be added to ring II, but their identity and position of attachment on ring II varies: paromomycin contains a 5-membered ribose and 6-membered xylose ring attached at position 5 of ring II whereas gentamicin contains a single 6-membered ring at position 6. All aminoglycosides are positively charged at neutral pH, making them attractive ligands for RNA recognition.

Antibiotic resistance and toxicity limit the utility of aminoglycoside antibiotics. The most widespread cause of aminoglycoside resistance is enzymatic modification of the antibiotic. Modification enzymes are plasmid encoded and thus represent a particularly promiscuous mechanism of resistance. The targets of the resistance enzymes are chemical groups in rings I and II, and the enzymes include acetyl transferases, phosphotransferases and adenylyl transferases (Shaw *et al.*, 1993). Enzymatic modification of the aminoglycoside binding site in the ribosome also yields resistance (Beauclerk and Cundliffe, 1987). A1408 or G1405 are targets of specific RNA methylases from bacteria that produce aminoglycosides. Resistance can arise through mutation of ribosomal RNA: a mispair at position 1409–1491 has been shown to infer high-level aminoglycoside resistance (DeStasio *et al.*, 1989). The functional importance of the A site, coupled with the large number of ribosomal RNA genes limits the importance of this mechanism. Toxicity to aminoglycosides arises from their known inhibition of eukaryotic ribosome function at 20-fold higher concentration (Palmer and Wilhelm, 1978). Sequence similarity between prokaryotic and eukaryotic ribosomes is the origin of this toxicity.

3 Defining a model oligonucleotide for structural studies

To determine the structural origin of aminoglycoside–rRNA recognition, we have used nuclear magnetic resonance spectroscopy. Unfortunately, the size limitation for NMR study (about 40 kD) is more than an order of magnitude smaller than the size of the ribosome. The local nature of the aminoglycoside binding site within 16 S ribosomal RNA suggests that a small domain of ribosomal RNA may recapitulate the antibiotic binding site, and be amenable to NMR. Purohit and Stern demonstrated that aminoglycosides bound specifically to a 64nt RNA that spanned the 30 S subunit A site (Purohit and Stern, 1994). We designed a shorter, 27nt model oligonucleotide that contains only the ribosomal A site (Recht *et al*., 1996) (Fig. 1c). The penultimate helix in 16 S ribosomal RNA was truncated to a 4 bp helix and capped by a tetraloop. The region of aminoglycoside interaction is adjacent to two G–C pairs, one of whose formation is confirmed by phylogenetic variation to A–U in mitochondrial ribosomal RNAs. We therefore modelled the A site as an asymmetric internal loop closed by two helical stems. Two additional G–C pairs were added to the upper stem to facilitate transcription by T7 RNA polymerase.

Before structure determination by NMR, the relevance of a model oligonucleotide to its biological counterpart must be proven. The binding of aminoglycoside antibiotics to the A-site model oligonucleotide was monitored using chemical probing methods (Recht *et al*., 1996). The RNA was modified in the presence and absence of aminoglycoside with the reagent dimethyl sulphate (DMS), which reacts with the N1 positions of adenine, N7 position of guanine and N3 position of cytosine. Protection of positions G1491N7, G1494N7, and A1408N1 are observed at 1 μM concentration of the aminoglycoside antibiotic paromomycin. More quantitative measurement yielded an approximate K_d of 0.2 μM. No footprint was observed when the antibiotic streptomycin, which binds to another region of 16 S ribosomal RNA, or the polycation spermine was added to the model oligonucleotide, indicating a specific interaction of paromomycin with the oligonucleotide. The same nucleotides are protected from reaction with DMS on binding of paromomycin to the 30 S subunit and the model oligonucleotide (Fig. 3). In summary, the biochemical experiments confirmed the specificity of paromomycin binding to the model oligonucleotide, and demonstrated the relevance of the model to ribosomal RNA.

	–Paromomycin		+Paromomycin	
Position	Oligo	30S	Oligo	30S
A1408	+	+	–	–
G1491	+	+	–	–
A1492	++	++	++	++
A1493	+	+	+	+
G1494	+	+	–	–

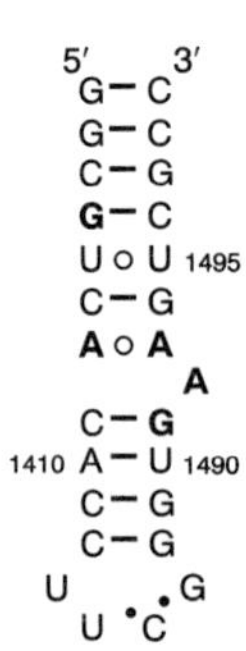
5′
3′
G–C
G–C
C–G
G–C
U○U 1495
C–G
A○A
A
C–G
1410 A–U 1490
C–G
C–G
U
G
U
C

Fig. 3 Comparison of chemical probing results on 30 S ribosomal subunits and the A-site model oligonucleotide. + indicates base modified; – indicates base not modified. Reactivity at the N7 of guanine and N1 of adenine to DMS is indicated by +; positions in the oligonucleotide are indicated in bold in the secondary structure. Both the oligonucleotide and the 16 S rRNA in 30 S subunits show increased reactivity of A1492 compared to A1493. Paromomycin protects the same set of nucleotides at 1 μM concentration in both the 30 S subunit and oligonucleotide (Moazed and Noller, 1987a; Woodcock *et al*., 1991).

The nucleotide determinants for aminoglycoside binding to ribosomal RNA were determined by means of mutant model oligonucleotides (Recht *et al.*, 1996) (Fig. 1d). The presence of a mispair at positions 1409–1491 is known to disrupt aminoglycoside binding to ribosomes, and induce aminoglycoside resistance. A similar mutation in the model oligonucleotide also disrupted binding, further confirming the relation of the model system to the ribosome. Further single nucleotide variants were constructed and probed using chemical reagents. The results of these studies are summarized in Fig. 4. A C1408–G1494 base pair is required for specific aminoglycoside binding, as is U1495, and the A1408–A1493 apposition. The presence of an additional nucleotide at position 1492, which maintains the asymmetry of the internal loop, is also required. Interestingly, U1406 can be changed to an A, with no effect on binding. The results of the mutational studies provide a biochemical context for the interpretation of the high resolution structure (see below).

4 RNA structure determination by NMR

Nuclear magnetic resonance spectroscopy provides high resolution structures of RNA and its complexes in solution. NMR measurements provide local distance and dihedral angle restraints between pairs of nuclei. Interproton distances are measured using the through-space dipolar coupling, or Nuclear Overhauser Effect (NOE) interaction, which depends on the interproton distance to the inverse sixth power; distances out to 5–6 Å can be semi-quantitatively determined. Dihedral angle restraints are determined from the angular dependence of the through-bond J-coupling term. Experimentally determined restraints are input into standard simulated annealing protocols, in which the restraint is treated as a pseudoenergy term. Molecular conformations are initially randomized, and conformational space is robustly sampled at high temperature. After cooling to low temperature, structures with low restraint violation energy and good geometry are considered converged.

Structural restraints are obtained from NMR data after individual resonances are assigned to a given nucleus. Two distinct classes of protons exist in RNA: exchangeable protons attached to electronegative atoms (N, O) and more numerous nonexchangeable protons attached to carbon. Modern multidimensional NMR methods allow complete proton assignments of RNAs of up to 35 nts. The oligonucleotide must be uniformly enriched in ^{13}C and ^{15}N. Biosynthetic methods for preparation of labelled nucleotide triphosphates allow ready enrichment of an RNA NMR sample (Batey *et al.*, 1995). Labelled RNA is also crucial for studies of RNA–ligand complexes, in which one component is isotopically labelled and the other is not. Isotope-filtered experiments allow selective NOE interactions to be observed only between RNA and ligand. These methodologies were essential to solve the structure of the aminoglycoside–A-site RNA complex.

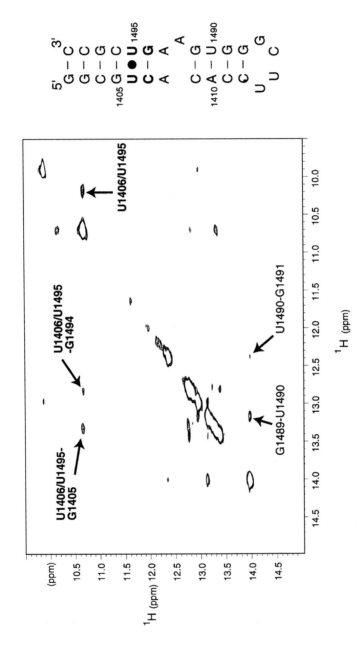

Fig. 4 NMR data indicate that the A-form RNA forms a closed internal loop structure. Shown is a NOESY experiment performed in H$_2$O at 5°C that shows NOEs between exchangeable imino protons. A strong NOE between U1406 and U1495 imino protons indicate formation of a U–U pair, which is followed by a C1407–G1494 pair. The mixing time for this experiment was 250 ms.

5 Conformation of the unbound A-site RNA

The conformation of the unbound form of the A-site oligonucleotide was determined using NMR. Two A-form helices, and the -UUCG- tetraloop form as predicted by the secondary structure. The internal loop is closed by formation of additional base pairs: a U1406–U1495 and a C1407–G1494 base pair (Fig. 4). The secondary structure of the RNA is readily determined by examination of the exchangeable imino proton spectrum; one imino proton resonance is expected for each Watson-Crick base pair. For the U–U pair, two imino proton resonances are observed (one from each uracil) and a strong NOE is observed between them, since they are less than 2.5 Å apart. The internal loop is thus closed to three adenosines, A1408, A1492 and A1493. The NMR data indicate that these adenosines do not form a tightly defined structure.

6 Structure of the A-site RNA–paromomycin complex

The binding of aminoglycoside antibiotics to the model oligonucleotide can be monitored by NMR spectroscopy. Figure 5 shows a titration of a 2 mM oligonucleotide solution with paromomycin; only the imino proton resonances are shown. As paromomycin is added, a set of peaks disappear, and a new set of resonances

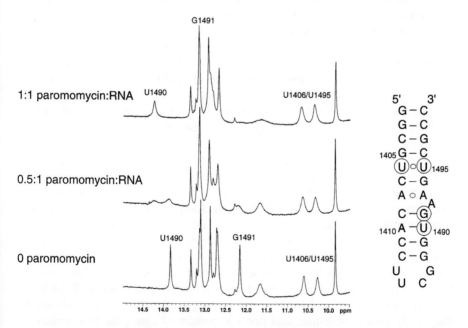

Fig. 5 Paromomycin forms a 1:1 complex with the A-site oligonucleotide. The imino proton spectrum for the RNA is shown at 0, 0.5:1 and 1:1 stoichiometry of paromomycin to RNA. A second set of peaks is observed, which indicates that the free and bound forms of the RNA are in slow exchange on the NMR time scale.

appear; this indicates that the free and bound forms of the RNA are in slow exchange on the NMR time scale. In addition, the titration demonstrates the 1:1 stoichiometry of the paromomycin–RNA complex. NMR titrations confirmed the results of biochemical studies. Variants with wild-type affinity for paromomycin behaved in a similar manner as wild-type oligonucleotide. Variants with perturbed affinity for paromomycin show distinctly different behaviour.

The structure of the 1:1 complex of paromomycin with the A site oligonucleotide was determined using NMR spectroscopy (Fourmy et al., 1996). A total of 392 NOE derived distance restraints, including 47 intermolecular restraints, was used to calculate the solution structure. The converged structures are well defined within the aminoglycoside binding site. The core region, which includes G1405 to A1410, U1490 to C1496 and all four rings of paromomycin has a root mean squared deviation of 0.61 Å among the 20 converged structures. The solution structure thus provides an atomic-level view of how aminoglycosides bind to their ribosomal target (Fig. 6).

Paromomycin binds in the major groove of the A-site RNA, within a binding pocket formed by the internal loop. The antibiotic adopts an L-shaped structure. Rings II, III and IV line the major groove, and ring I fits into a specific pocket formed by an A1408–A1493 base pair and the bulged adenosine A1492. Rings I and II are the best defined portions of paromomycin, and direct specific interaction with the RNA, consistent with the conservation of these rings in different aminoglycosides that target the ribosomal A-site. Rings III and IV vary among aminoglycosides, and these rings are more disordered in the NMR structure.

The A-site RNA adopts a similar conformation in the complex as in the unbound form. The internal loop is closed by the U1406–U1495 and C1407–G1494 base pairs. The conformation of A1408, A1492 and A1493 are fixed compared to the free form. A1408 and A1492 form a base pair, with the Watson-Crick face of A1408 interacting with the Hoogsteen face of A1493; A1492 is displaced towards the major groove, and the phosphodiester backbone between G1494 and G1491 forms the lip of the drug binding pocket.

Rings I and II of paromomycin direct specific interaction with the A-site RNA (Fig. 7). The two functionally important amino groups in ring II form hydrogen bonds with the N7 of G1494 and the O4 of U1406. The formation of these hydrogen bonds is supported by the functional importance of G1494 and U1495 for aminoglycoside binding. Ring I sits within the binding pocket formed by the A1408–A1493 base pair and the bulged nucleotide A1492. Chemical groups in ring I contact the phosphodiester backbone between G1494 and G1491, including the conserved hydrogen bond donor at the 6' position that contacts the phosphate between G1494 and A1493. No direct hydrogen bonding contacts to RNA bases are made by ring I. The bottom of the binding pocket for ring I is formed by the C1409–G1491 base pair, and ring I is positioned above the base moiety of G1491. The elements of the binding pocket, A1408–A1493 base pair, a bulged nucleotide at 1492 and a Watson-Crick base pair at position 1409–1491, are essential for high affinity paromomycin binding (Recht et al., 1996).

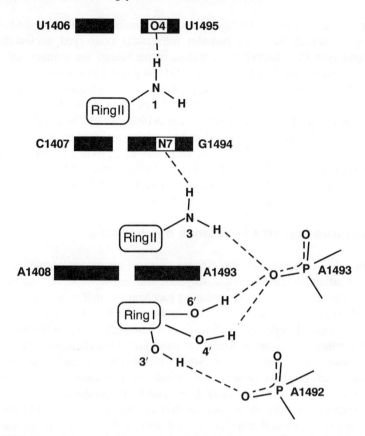

Fig. 7 Schematic of intermolecular contacts between paromomycin and the A-site RNA oligonucleotide.

Rings III and IV are less crucial for the antibiotic action of aminoglycosides. Rings III and IV can be deleted or changed and antibiotic action is maintained, albeit at a higher effective concentration. Rings III and IV are less well defined in the NMR structure of the paromomycin complex, and this disorder likely reflects the dynamic nature of these rings. Rings III and IV are located in the major groove of the lower helical stem near base pairs C1409–G1491 and A1410–U1490. These rings make primarily electrostatic contacts with the phosphodiester backbone of the RNA helix. The lack of sequence-specific contacts is supported by our mutational analyses of base pairs in the lower stem.

7 Prokaryotic specificity of aminoglycoside antibiotics

Aminoglycoside antibiotics cause miscoding and inhibit translation at lower concentration for prokaryotic ribosomes than eukaryotic ribosomes. These differences represent differences in the binding affinities of aminoglycosides to prokaryotic v. eukaryotic ribosomes. The aminoglycoside binding site is made up of two halves: one contains universally conserved nucleotides and the second contains variable nucleotides. In particular, the contacts of ring II are to universally conserved nucleotides G1494 and U1495. However, A1408 is replaced by a G in all eukaryotic sequences. This single mutation reduces the binding affinity of paromomycin to the model oligonucleotide by 15-fold (Recht *et al.*, 1996). The specific geometry of the A1408–A1493 base pair is required to form the binding pocket for ring I and cannot be replaced by a G1408–A1493 pair of similar geometry. In addition, all higher eukaryotic ribosomes contain a mispair at the 1409–1491 position, which forms the bottom of the binding pocket. The combination of the G1408 substitution and 1409–1491 mispair explains the specific action of aminoglycosides on prokaryotic organisms.

8 Structure explains resistance mechanisms

Resistance has compromised the efficacy of most antibiotics. Aminoglycosides have been administered for almost 40 years, and resistance is widespread. Antibiotic resistance can arise from transport mechanisms that pump antibiotics out of the cell, from enzymatic modification of the antibiotic, or from modification of the cellular target. Enzymatic modification of aminoglycosides is the most prevalent mechanism of resistance. These enzymes are plasmid encoded and are thus readily transmitted between strains. Aminoglycoside modification enzymes target chemical groups on rings I and II, and include acetyl transferases, phosphotransferases and adenylyl transferases (Fig. 8). Antibiotic-producing organisms must protect themselves from their own antibiotics, and aminoglycoside-producing organisms encode methyl transferases that specifically modify ribosomal RNA to yield resistance (Fig. 9). These enzymes include an A1408 N1 methylase

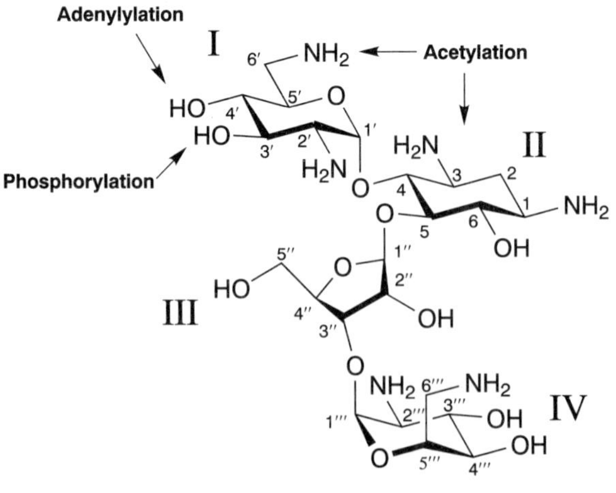

Fig. 8 Targets of aminoglycoside modification enzymes on neomycin.

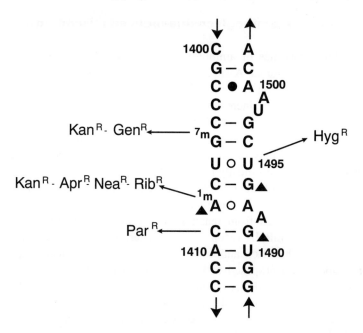

Fig. 9 Modifications of 16 S ribosomal RNA that yield antibiotic resistance. Sites of methylation, and the spectrum of resistance that results, are indicated for G1405 and A1408. RNA mutations at 1409–1491 and 1495 that give resistance are also shown.

that gives broad spectrum aminoglycoside resistance and G1405 N7 methylase that gives resistance only to gentamicin and kanamycin. Several ribosomal RNA mutations have been isolated that give aminoglycoside resistance, including the 1409–1491 mispair, and U1495 to C mutation that yields hygromycin resistance. These mutations are rare, as there are often multiple copies of ribosomal RNA genes in bacteria.

The paromomycin–A-site RNA structure explains the origins of these resistance mechanisms. The targets of aminoglycoside modification enzymes are in regions of close RNA–aminoglycoside contact, such that acetylation, phosphorylation and adenylylation would disrupt specific hydrogen bonding and electrostatic contacts. In contrast, the ring III-amino group is not a target of acetyl transferases, since the amide NH can still hydrogen bond to U1495 O4 and exit out the top of the major groove with no steric clash. The RNA modification enzymes also target critical elements for antibiotic binding. Methylation of A1408 N1 would disrupt the A1408–A1493 pair that forms the binding pocket for ring I. The N7 of G1405 is distant from paromomycin in the NMR spectrum. However, methylation of G1405 causes resistance only to gentamicin and kanamycin, whose ring III is attached to ring II at position 6. If these antibiotics bind in a similar way to the A site, then their ring III would be in closer proximity to the N7 of G1405.

9   Mechanism of aminoglycoside-induced miscoding

The miscoding properties of aminoglycosides provide insight into the natural functions of the ribosomal A-site. Binding of the aminoglycoside antibiotics decreases the dissociation rate constant for A-site bound tRNAs (Karimi and Ehrenberg, 1994), and the increased affinity of the A-site is consistent with the miscoding induced by aminoglycosides. Since ribosome contacts with the A-site tRNA–mRNA complex are an essential feature of models for translational accuracy, aminoglycosides may favour these contacts. In the paromomycin complex, A1492 and A1493 are specifically positioned with their N1s pointing in the minor groove. In the absence of an antibiotic, the adenines do not adopt a fixed conformation. The N1 positions of these two universally conserved adenines A1492 and A1493 are protected from reaction with dimethyl sulphate upon binding of A-site tRNA and mRNA (Moazed and Noller, 1990). Based on the conformational change induced on antibiotic binding, and the physical measurements mentioned above, we proposed that the bound conformation represents a high affinity conformation for tRNA binding in the A-site (Fourmy et al., 1996). The N1 positions of A1492 and A1493 are positioned such that they can hydrogen bond with two 2'-OH groups in the RNA helix formed by the messenger RNA in the codon–anticodon complex.

The proposed mechanism presents specific details for the contribution of the ribosome to the accuracy of protein synthesis. Contacts between the ribosome and the codon–anticodon complex must occur in a sequence independent manner. All cognate codon–anticodon complexes form at least two Watson-Crick pairs in an A-form geometry, and thus should present two 2'-OH in a similar orientation. A non-cognate complex would not present the 2'-OH in the proper orientation. In the absence of an aminoglycoside antibiotic, the contacts between A1492/1493 and the two 2'-OH would drive the conformational change of the A-site rRNA. The small free energy cost of the conformational change is overcome by the binding of an aminoglycoside antibiotic in the A-site. The proposed contacts are consistent with limited biochemical data. Changes of universally conserved A1492 and 1493 yield lethal phenotypes and substitutions of 2'-OH with deoxy nucleotides in the A-site codon decrease the binding affinity of the cognate tRNA.

10   Conclusions

The structure of the paromomycin–A-site RNA complex explains a variety of biochemical and pharmacological results on aminoglycoside antibiotics and translation. The results of this work demonstrate that local domains of the ribosome can be recapitulated as model oligonucleotides. The structure illustrates how an RNA-directed therapeutic agent recognizes its target, and represents a starting point for drug design. From analysis of the structure arise testable hypotheses of ribosome function.

Acknowledgements

We would like to thank Prof. Harry Noller and members of his research group for discussion of ribosome structure and function. The work was supported by grants from the Packard Foundation, Deafness Research Foundation, National Institute of Health (GM51266-01A1), and Lucille P. Markey Charitable Trust.

References

Batey, R. T., Battiste, J. L. and Williamson, J. R. (1995). *Methods Enzymology* **261**, 300–323.

Beauclerk, A. A. and Cundliffe, E. (1987). *J. Mol. Biol.* **193**, 661–671.

Davies, J. and Davis, B. D. (1968). *J. Biol. Chem.* **243**, 3312–3316.

Davies, J., Gorini, L. and Davis, B. D. (1965). *Mol. Pharmacol.* **1**, 93–106.

Davies, J., Jones, D. S. and Khorana, H. G. (1966). *J. Mol. Biol.* **18**, 48–57.

DeStasio, E. A., Moazed, D., Noller, H. and Dahlberg, A. E. (1989). *EMBO J.* **8**, 1213–1216.

Edelmann, P. and Gallant, J. (1977). *Cell* **10**, 131–137.

Fourmy, D., Recht, M., Blanchard, S. D. and Puglisi, J. D. (1996). *Science* **274**, 1367–1371.

Gale, E. F., Cundliffe, E., Reynbolds, P. E., Richmond, M. H. and Waring, M. J., eds (1981). In *The Molecular Basis of Antibiotic Action*. John Wiley & Sons, London.

Gutell, R. R. (1994). *Nucleic Acids Res.* **22**, 3502–3507.

Karimi, R. and Ehrenberg, M. (1994). *Eur. J. Biochem.* **226**, 355–360.

Kurland, C. G. (1992). *Ann. Rev. Genet.* **26**, 29–50.

Moazed, D. and Noller, H. F. (1987a). *Biochimie* **69**, 879–884.

Moazed, D. and Noller, H. F. (1987b). *Nature* **327**, 389–394.

Moazed, D. and Noller, H. F. (1990). *J. Mol. Biol.* **211**, 135–145.

Noller, H. F. (1991). *Ann. Rev. Biochem.* **60**, 191–227.

Noller, H. F., Hoffarth, V. and Zimniak, L. (1992). *Science* **256**, 1416–1419.

Palmer, E. and Wilhelm, J. M. (1978). *Cell* **13**, 329–334.

Powers, T. and Noller, H. F. (1995). *RNA* **1**, 194–209.

Purohit, P. and Stern, S. (1994). *Nature* **370**, 659–662.

Recht, M. I., Fourmy, D., Blanchard, S. C., Dahlquist, K. D. and Puglisi, J. D. (1996). *J. Mol. Biol.* **262**, 421–436.

Shaw, K. J., Rather, P. N., Hare, R. S. and Miller, G. H. (1993). *Microbiol. Rev.* **57**, 138–163.

Woodcock, J., Moazed, D., Cannon,. M., Davies, J. and Noller, H. F. (1991). *EMBO J.* **10**, 3099–3103.

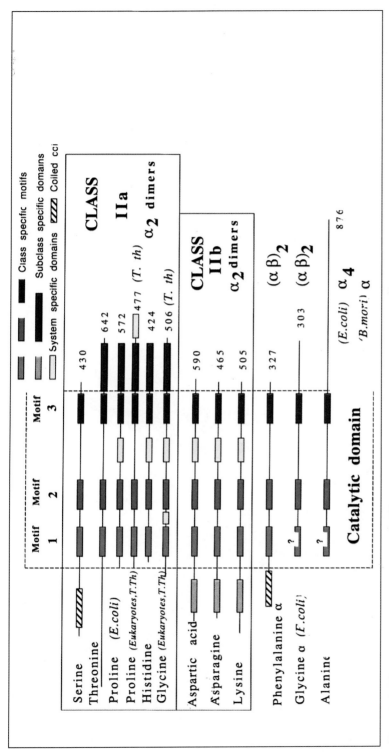

Chapter 4, Fig. 1 Schematic diagram showing the modular structure of class II aminoacyl-tRNA synthetases. The polypeptide chain length refers to the *E. coli* enzyme except where otherwise stated. The class II specific motifs in the catalytic domain are shown in green (motif 1), blue (motif 2) and red (motif 3). Subclass conserved anticodon binding domains are mauve (class IIa) or orange (class IIb). System conserved domains (yellow) are of variable size and tertiary fold. PheRS and SerRS have N-terminal coiled-coils ('helical arms', shown hatched). Note the two distinct forms of GlyRS and ProRS (see text).

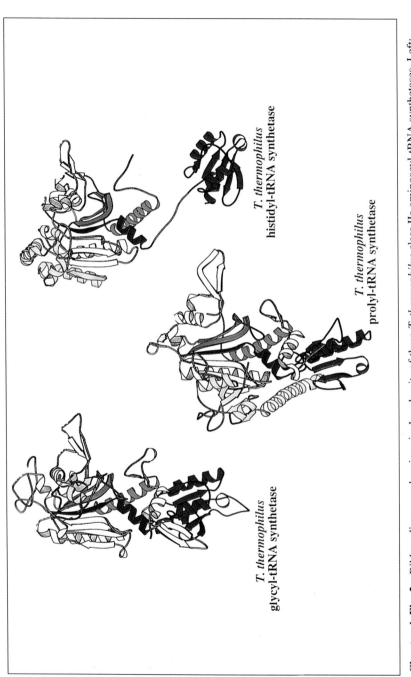

Chapter 4, Fig. 2   Ribbon diagrams showing single subunits of three *T. thermophilus* class IIa aminoacyl-tRNA synthetases. Left: glycyl-tRNA synthetase (Logan *et al.*, 1995), centre: prolyl-tRNA synthetase (Cusack *et al.*, unpublished), right: histidyl-tRNA synthetase (Åberg *et al.*, 1997). Colour scheme corresponds with Fig. 1.

T. thermophilus histidyl-tRNA synthetase

T. thermophilus prolyl-tRNA synthetase

T. thermophilus glycyl-tRNA synthetase

T. thermophilus
asparaginyl-tRNA synthetase

T. thermophilus
lysyl-tRNA synthetase

yeast
aspartyl-tRNA synthetase

Chapter 4, Fig. 3 Ribbon diagrams showing single subunits of the three class IIb aminoacyl-tRNA synthetases. Left: *T. thermophilus* asparaginyl-tRNA synthetase (Berthet-Colominas *et al.*, unpublished), cenre: *T. thermophilus* lysyl-tRNA synthetase (Cusack *et al.*, unpublished), right: yeast aspartyl-tRNA synthetase (Ruff *et al.*, 1991). Colour scheme corresponds with Fig. 1.

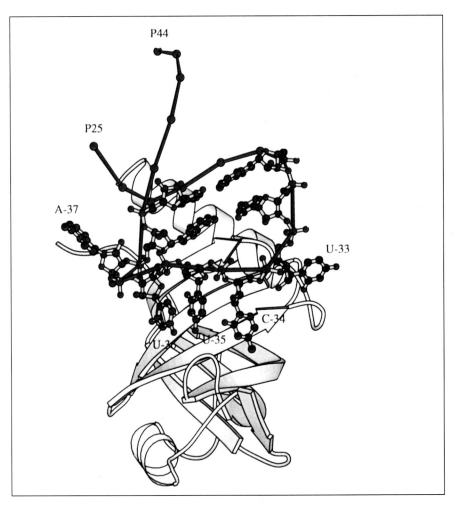

Chapter 4, Fig. 4 The interaction of the anticodon stem-loop of tRNA^lys with the N-terminal β-barrel domain (OB-fold) of *T. thermophilus* lysyl-tRNA synthetase. The three anticodon bases make specific interactions with the synthetase as shown in detail in Fig. 5. The N-terminal helix specific to LysRS (see text) is at the top and back of the OB-fold module.

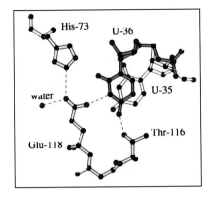

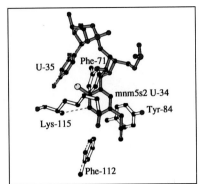

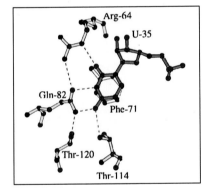

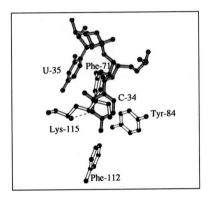

Chapter 4, Fig. 5   Specific interactions of the three anticodon bases of tRNA[lys] with *T. thermophilus* lysyl-tRNA synthetase; (**a**) U-36; (**b**) U-35, (**c**) mnm^{5}s^{2}U-34 as in *E. coli* tRNA[lys]; (**d**) C-34 as in *T-thermophilus* LysRS. Reproduced with permission from *Cusack et al.*, 1996b.

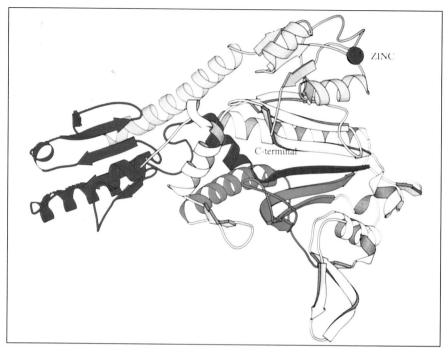

Chapter 4, Fig. 6   Ribbon diagram of a single subunit *T. thermophilus* prolyl-tRNA synthetase showing the position of the zinc atom and the extreme C-terminus in the active site. The colouring of domains corresponds to that in Figs 1 and 2.

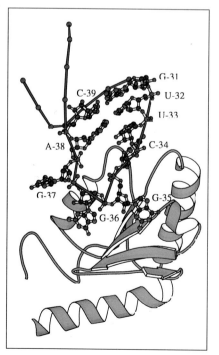

Chapter 4, Fig. 7   The interaction of the anticodon stem-loop of tRNApro with the C-terminal α/β domain of *T. thermophilus* prolyl-tRNA synthetase. Only the second two bases of the anticodon, G-35 and G-36, make specific interactions with the synthetase.

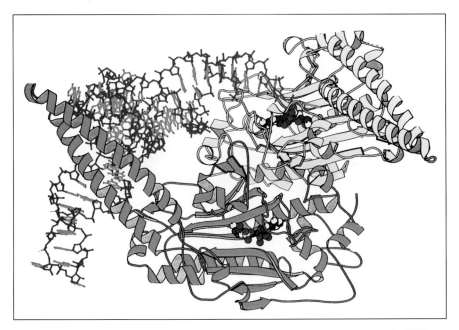

Chapter 4, Fig. 8   Overall view of the ternary complex of *T. thermophilus* seryl-tRNA synthetase with tRNA^ser^(GGA) and a synthetic analogue of seryl-adenylate (Ser-AMS). The synthetase monomer 1 is in yellow and monomer 2 in blue. The tRNA backbone is in red and bases in green. The tRNA is viewed looking down the anticodon stem which is not ordered in the crystal structure and not included in the figure. The long variable arm of the tRNA crosses perpendicularly the helical arm of monomer 2 and emerges at the bottom left of the figure. In each active site, the Ser-AMS molecule is represented by spheres. Experimental electron density is poor or lacking for two thirds of the helical arm of monomer 1, the 3' terminal C-75 -76 of the tRNA and the extreme end of the tRNA long variable arm. These regions have been manually modelled. Reproduced with permission from Cusack *et al.*, 1996a.

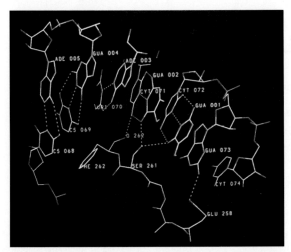

Chapter 4, Fig. 9   View of the motif 2 loop of seryl-tRNA synthetase interacting in the major groove of the acceptor stem of tRNA^ser^. Reproduced with permission from Cusack *et al.*, 1996a.

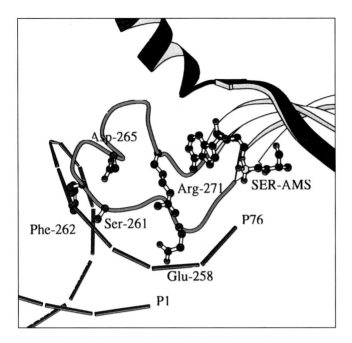

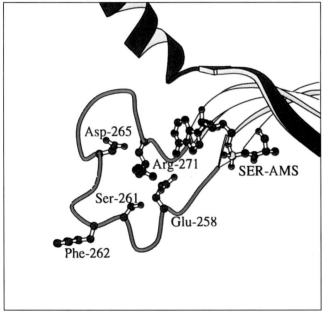

Chapter 4, Fig. 10 Two active conformations of the motif 2 loop (green) of seryl-tRNA synthetase. Top: 'T-conformation', observed in the ternary complex with tRNA[ser] and a non-hydrolyzable analogue of seryl-adenylate (denoted SER-AMS). Bottom: 'A-conformation', observed in the presence of ATP or seryl-adenylate alone. Reproduced with permission from Cusack *et al.*, 1996a.

8

Protein–RNA Recognition in a Highly Conserved Ribosomal Domain Targeted by Thiazole Antibiotics*

DAVID E. DRAPER

Department of Chemistry, Johns Hopkins University, Baltimore, MD 21218, USA

Abstract

A number of protein–RNA interactions in the ribosome have been highly conserved during evolution and must be among the most ancient examples of protein–nucleic acid recognition in the cell. Among them is L11 which recognizes a 58 nucleotide rRNA fragment. The structure of a 76 residue L11 RNA-binding domain, both free in solution and bound to its RNA target, has been solved by NMR methods, and the RNA-binding surface defined by site-directed mutagenesis and NMR. Two main features contribute to recognition: an α-helix and a large, flexible loop that adopts a specific conformation upon binding. The protein structure is remarkably similar to that of the homeodomain class of DNA-binding proteins. The similarity extends to the use of the same helix region for nucleic acid recognition, and the strategy of a flexible loop wrapping around the nucleic acid is reminiscent of 'wings' or loops found in many helix-turn-helix DNA binding proteins. Since L11 is considerably more ancient than homeodomain proteins, L11 may have been an evolutionary ancestor of homeodomains. Alternatively, the homeodomain protein fold may be a particularly versatile motif for nucleic acid recognition that has arisen a number of times during evolution.

The RNA target of L11 has a very stable tertiary structure that folds around Mg^{2+} and NH_4^+ in preference to other ions. L11 stabilizes the tertiary structure further and probably contacts an internal loop that is a key part of the tertiary

---

*Figure 1 for this chapter appears in the colour plate section between p. 144 and p. 145.

THE MANY FACES OF RNA
ISBN 0-12-233210-5

structure. An antibiotic, thiostrepton, recognizes the same tertiary structure and interacts cooperatively with the N-terminus of L11.

Phylogenetic analysis has been very helpful in suggesting key components of the RNA tertiary structure and the L11 recognition surface: the most highly conserved residues (either protein or RNA) tend to be required either for folding or for recognition. However, the phylogenetic record also suggests that bacterial, archaeal and eukaryotic complexes use slightly different strategies for recognition. Since thiostrepton discriminates between bacterial and eukaryotic ribosomes, these small differences may be appropriate targets for drug design.

1 Introduction

Ribosomes are ubiquitous to living organisms and must have evolved to their present form very early in biotic evolution. Olsen and Woese and others some time ago called attention to the fact that all ribosomal RNA sequences can be aligned, and this is now a powerful tool for establishing phylogenetic relationships (Olsen and Woese, 1993). The spectacular conservation of ribosomal RNA sequences and secondary structures originally inspired arguments that ribosomal RNAs must be functionally more important than ribosomal proteins. However, now that more ribosomal proteins have been sequenced it is clear that many of them are as highly conserved as the RNA. The ribosome appears to have an RNA–protein 'core' that has remained unchanged since the divergence of the main phylogenetic domains.

The ribosome is, in effect, a collection of ancient RNA–protein complexes that is a wonderful resource for the student of RNA–protein interactions. During the long evolutionary history of the ribosome, some of its proteins and RNA domains may have been recruited for other purposes in the cell: one might therefore expect the ribosome to reflect fundamental themes in nucleic acid–protein interactions. Another consequence of the antiquity of the ribosome and its central importance to cellular function is that it has been a prime target for antibiotic evolution. How these smaller ligands have been able to take advantage of subtle differences between classes of ribosomes (for example, bacterial vs. eukaryotic) or co-evolve with resistance mechanisms is an interesting topic in itself, with considerable practical importance for design of new drugs. A last advantage of studying ribosomal protein–RNA complexes is the large database of sequences that is now available. Conservation and variation among both RNA and protein sequences provide valuable clues to the folding of both components and help identify functionally important residues.

This review describes the current structural and thermodynamic picture of ribosomal protein L11, its interaction with RNA, and the thiazole-containing antibiotics which target this complex. Cundliffe's laboratory first established that a very stable ternary complex between a small fragment of 23 S rRNA, protein L11, and the antibiotic thiostrepton could be formed (Thompson *et al.*, 1979). *In vivo* resistance to thiostrepton is conferred either by 2′-O-methylation of a nucleotide within the rRNA binding site (the natural resistance mechanism of the

Streptomyces strain which synthesizes thiostrepton) (Thompson *et al*., 1982), by mutations in the large subunit rRNA (Hummel and Böck, 1987), or by mutations within L11 (Spedding and Cundliffe, 1984). Ribosomes lacking L11 are functional, but adding back L11 stimulates elongation factor G-dependent GTP hydrolysis and increases translation rates by about a factor of two (Stark and Cundliffe, 1979).

For the purposes of this review, it is important to note that the L11–RNA interaction has been functionally conserved. A number of experiments have tested an L11 homologue from one species with RNA from another; specific recognition has invariably been detected (Thompson *et al*., 1979; Beauclerk *et al*., 1985; El-Baradi *et al*., 1987). Substitution of the *E. coli* rRNA sequence binding L11 into yeast large subunit rRNA results in functional ribosomes (Musters *et al*., 1991), and the reciprocal experiment has the same result (Thompson *et al*., 1993). We therefore expect that important contacts between the protein and RNA have not varied during evolution. The situation is different with thiostrepton: bacterial ribosomes are generally sensitive to the antibiotic, while eukaryotic ribosomes are resistant and archaebacterial ribosomes show a range of sensitivities (Amils *et al*., 1990). Thiostrepton must take advantage of subtle differences between bacterial and eukaryotic L11–RNA complexes to discriminate between them.

2 Structure of the RNA binding domain of ribosomal protein L11

Structural work on ribosomal proteins made little progress for many years due to the instability of *E. coli* proteins apart from the ribosome. This problem was first circumvented by White and Ramakrishnan, who found that some *Bacillus stearothermophilus* (Bst) ribosomal proteins crystallized readily (Ramakrishnan and Gerchman, 1991; Ramakrishnan and White, 1992). Since then structures of a number of ribosomal proteins from thermophiles have been solved by NMR or crystallography (Liljas and Garber, 1995). The Bst homologue of L11 was cloned and overexpressed, but was not soluble to the concentrations needed for structural studies (Xing and Draper, 1995). However, trypsin digestion of an L11–RNA complex yielded a 75 residue C-terminal fragment with identical RNA-binding properties as the intact protein (Xing and Draper, 1996). This fragment, called L11-C76 when cloned and overexpressed, had excellent solubility properties and its structure was readily determined by NMR (Xing *et al*., 1997).

The protein backbone is shown in Fig. 1a. Three α-helices form the hydrophobic core of L11-C76, and a short parallel β structure joins the ends of helices II and III. The large loop between helices I and II is completely disordered: a proline is rapidly switching between *cis* and *trans* conformations, and ^{15}N relaxation times indicate conformational motions on a very rapid time scale (Markus *et al*., 1997).

Having obtained this structure, we naturally wondered what part of the protein is responsible for RNA recognition. Our thinking in this regard was aided by startling similarities between the L11-C76 structure and classes of DNA-binding proteins. We first noticed that the three α-helices of L11-C76 align very well with the α-helices of homeodomain proteins, with an r.m.s. deviation of 1.2 Å (an example homeodomain protein, engrailed, is shown alongside L11-C76 in Fig. 1). Homeodomains are a frequent motif among transcription factors regulating development in higher organisms. Their mechanism of DNA recognition is well known from a number of homeodomain–DNA complex crystal structures (Kissinger *et al.*, 1990; Klemm *et al.*, 1994; Hirsch and Aggarwal, 1995; Li *et al.*, 1995): helix III lies in the DNA major groove, making hydrogen bonding contacts with bases, and the N-terminus wraps around the DNA into the minor groove. This suggested to us that L11 helix III might similarly be involved in RNA recognition. We also noticed that solvent-exposed residues in helix III are among the most highly conserved residues in the protein (Fig. 2); since L11 is interchangeable between organisms, we would expect the RNA-binding surface to contain conserved residues. Site-directed mutagenesis of conserved surface residues turned up several helix III substitutions that dramatically weakened RNA binding affinity (Fig. 2) (Xing *et al.*, 1997), supporting the idea that helix III is involved in RNA recognition.

Homeodomain proteins are part of the larger helix-turn-helix class of DNA binding proteins. These proteins all have two α-helices, equivalent to helices II and III of homeodomains and L11-C76. The second (C-terminal) helix is responsible for DNA recognition, though different classes of helix-turn-helix proteins position the helix differently on the DNA surface (Treisman *et al.*, 1992).

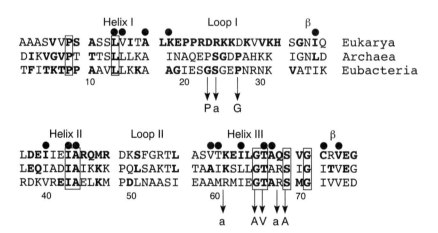

Fig. 2 Conservation of residues within the L11 RNA binding domain. Representative sequences from three phylogenetic domains are shown (eukarya: *Saccharomyces cerevisiae*; archaea: *Sulfolobus acidocaldarius*; bacteria: *Bacillus stearothermophilus*). Numbering is for the L11-C76 protein (Fig. 1). Residues that are conserved in >80% of available sequences within a phylogenetic domain are displayed in boldface type, and residues conserved in all three domains are boxed. Dots above residues indicate involvement in the protein hydrophobic core. Arrows indicate mutations that have been made in L11-C76; mutations weakening RNA binding by more than 10 fold are in upper case.

A second feature of L11-C76 that is reminiscent of DNA binding proteins is its disordered loop, residues Ala 18–Lys 31. Several classes of transcription factors have the α-helix core of homeodomains but in addition have one or two large loops (sometimes called 'wings') that wrap around the DNA. Figure 1 shows two of these, the H5 histone globular domain and the ETS domain of transcription factor PU.1. A similar protein, Mu transposase, has been studied by NMR both free in solution and bound to its DNA target. It is clear from ^{15}N-relaxation studies that a loop in this protein is quite flexible in the absence of DNA, but is fixed in a specific conformation upon binding to DNA (Clubb *et al.*, 1996). Similar ^{15}N-relaxation studies of L11-C76 either free in solution or bound to RNA (Markus *et al.*, 1997) show that the loop becomes as ordered as the rest of the protein upon binding to the RNA. Some mutations at conserved residues within this loop also strongly reduce RNA binding affinity (Fig. 2) (Xing *et al.*, 1997). It is not clear from these results whether the mutations prevent correct folding of the flexible loop or directly alter contacts with the RNA. Recent determination of the structure of L11-C76 when bound to RNA confirm that the loop directly contacts the RNA surface (A. Hinck, M. Markus, S. Huang, D. Draper, and D. Torchia, unpublished observations).

The flexible loop, which has several residues that are well conserved among bacteria, has been substituted with an entirely different sequence in eukaryotes (Fig. 2). It is especially remarkable that eight of the 16 residues in the eukaryotic loop are basic, compared to two to four basic residues (out of 14) in bacterial sequences. Since the yeast and *E. coli* L11 homologues are interchangeable (El-Baradi *et al.*, 1987), it is interesting to ask what kind of compensation accounts for this dramatic difference in a part of the protein directly contacting the RNA. One possibility is that the eukaryotic sequence is able to make exactly the same RNA contacts as the bacterial sequence, but uses a much different folding strategy to position hydrogen bonding groups on the RNA surface. Alternatively, the two different loop sequences could be contacting different, but equally well conserved, regions of the RNA. In either case, the loop provides an interesting target for antibiotics specific for bacteria over eukaryotes.

3   An RNA tertiary structure dependent on Mg^{2+} and NH_4^+ ions

L11 recognizes an RNA domain with a minimum of 58 nucleotides (Ryan and Draper, 1989), (Fig. 3). The sequence is strongly conserved. The four base pairs at the 5'-3' termini form the longest standard Watson-Crick helix in the domain: mismatches or internal loops interrupt the other stems, and two conserved Watson-Crick pairs have no neighbours in the secondary structure. It thus seemed very likely that additional tertiary or non-canonical interactions stabilize the conserved pairing in this RNA. A melting profile of the *E. coli* sequence is shown in Fig. 4. By comparing this profile with the melting profiles of smaller RNAs reproducing the two hairpins and with mutations substituting different base pairs, we have been able to conclude that all the Watson-Crick secondary structure melts in

D. E. Draper

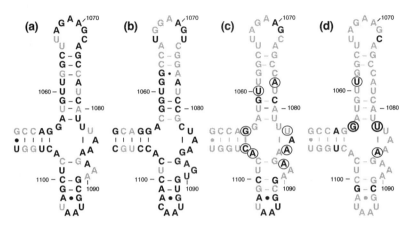

Fig. 3 Conservation of the large subunit ribosomal RNA domain binding L11; (a) *Escherichia coli* sequence 1051–1108. Bases shown in black are conserved among a set of 65 bacterial sequences with no more than two exceptions; (b) Yeast (*Saccharomyces cerevisiae*) sequence 1225–1282. Conservation of bases are indicated as in panel A (from a set of 49 sequences); (c) positions conserved as the same base in both eukaryotic and bacterial phylogenetic domains are shown in black. Circled bases are positions where mutations prevent tertiary structure formation (Lu and Draper, 1994); (d) bases that are conserved in bacteria and eukaryotes, but as different bases, are in black. The four circled bases must all be substituted with the eukaryotic sequences to maintain a stable tertiary structure.

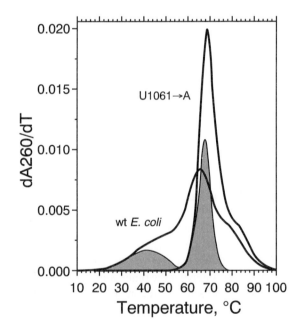

Fig. 4 Unfolding of a set of RNA tertiary interactions. The 58 nucleotide *E. coli* sequence shown in Fig. 3a was melted in buffer containing $MgSO_4$ and KCl (Lu and Draper, 1994). Both the wild type and a $U1061 \rightarrow A$ mutation are shown. The grey transitions represent the melting of the tertiary structure, as estimated by a comparison of melting profiles at two wavelengths (Lu and Draper, 1994).

transitions above $\sim$55°C (Laing and Draper, 1994). This leaves the first transition (at $\sim$40°C) as potential tertiary unfolding, and in fact it has several unusual properties:

(i) The hyperchromicity at 280 nm is nearly zero.

(ii) The transition appears only when Mg^{2+} is present, and Mg^{2+} is much more effective at inducing the folded structure than other alkaline earth ions. A highly cooperative uptake of divalent ions is characteristic of the folding of larger RNAs, such as group I introns (Laggerbauer et al., 1994), into their active tertiary structures. Transfer RNA is greatly stabilized by Mg^{2+} but does not require the ion for its tertiary structure to fold (Cole et al., 1972), and there is little difference in the effectiveness of different divalent ions. Both the requirement for Mg^{2+} and the specificity for this ion are therefore unusual.

(iii) The transition appears only when NH_4^+ or K^+ is present, with NH_4^+ stabilizing the structure somewhat more effectively than K^+ (Wang et al., 1993). This effect is completely unexpected. The only other instance of a nucleic acid having substantial specificity for monovalent ions is the G-quartet structure, in which ions occupy a channel formed by four nucleic acid strands (Williamson et al., 1989).

The specificity of this structure for mono- and di-valent ions suggests that the RNA adopts a precise tertiary structure that extensively coordinates or hydrogen bonds at least two ions. A protein analogy to this RNA structure might be a zinc finger, in which the zinc ion is buried within the protein and induces folding of a structure that would otherwise be unstable (Berg, 1990).

The sequence conservation of this RNA domain should provide some clues to the nucleotides important for folding its tertiary structure. Figure 3c shows those nucleotides that are conserved between bacterial and eukaryotic sequences. Nineteen of the 58 nucleotides are invariant or nearly so. Mutations at 10 of these sites have been tested in melting experiments, using the hyperchromicities at 260 and 280 nm and the effects of ion substitutions on the melting profiles as assays for tertiary structure formation (Lu and Draper, 1994). Nine of the 10 mutants tested eliminate the tertiary structure (circled bases in Fig. 3c); the exception is A1084 $\rightarrow$ U, which only destabilizes the structure by about 10 deg.

Some of the nucleotides are highly conserved in both bacteria and eukarya, but as different bases (Fig. 3d). Particularly intriguing is the interchange of U1082–A1086 in bacteria with C1082–G1086 in eukaryotes, which is evidence for Watson-Crick pairing between these positions. We found, however, that mutation of E. coli RNA to contain the eukaryotic C1082–G1086 resulted in an unusual melting profile: the tertiary structure was not present, but the secondary structure appeared to be stabilized (Lu and Draper, 1994). The change was consistent with formation of a G1056–C1082 pair forming in preference to C1082–G1086. Since the bacterial G1056 is almost always substituted by A in eukaryotes, we reasoned that this change is needed to prevent incorrect hydrogen bonding. The triple mutant, A1056-C1082-G1086, does have detectable tertiary structure in melting experiments, though it is much less stable than the wild type sequence. Inclusion of a fourth mutation, U1061 → G, now restored the tertiary structure to the same stability as the wild type sequence. This mutation by itself stabilizes the tertiary structure dramatically and evidently counters the weakened structure made by A1056-C1082-G1086. It is interesting to note that U1061 → A stabilizes the tertiary structure even more effectively than U1061 → G, increasing the tertiary structure T_m by 25 deg, but is rarely found in nature. It is possible that the tertiary structure can be made too stable for ribosomes to function, and that evolution has selected sequences with the tertiary structure adjusted to a moderate stability.

Our experiments with different cations and mutants have suggested that none of the RNA secondary base pairs may melt before the set of Mg^{2+} and NH_4^+-stabilized tertiary interactions. This is dramatically demonstrated by comparing melting profiles of the wild type *E. coli* rRNA fragment with the U1061 → A mutation (Fig. 4): stabilization of the tertiary structure has pushed secondary structure unfolding to higher temperatures. Obligatory unfolding of tertiary interactions before secondary structure has also been observed in tRNA (Stein and Crothers, 1976), and is a natural consequence of the fact that tertiary interactions essentially crosslink the junction of cloverleaf helices. The rRNA fragment tertiary structure probably does the same for its three-helix junction.

4   RNA features required for protein recognition

RNA tertiary structure can be an indirect determinant of protein recognition, in the sense that the tertiary structure may create an optimum three-dimensional configuration of the nucleotides contacting protein. Many aminoacyl-tRNA synthetases, for instance, primarily contact the tRNA anticodon loop and the acceptor stem. These two recognition features do not directly participate in tertiary interactions, but the tertiary structure does hold them in a specific orientation with respect to each other. A stable tertiary structure therefore promotes protein recognition (Puglisi *et al.*, 1993). Since the minimum size RNA fragment recognizing L11 has an extensive tertiary structure, it might be expected that the tertiary structure will be required for L11 recognition.

We have two arguments that tertiary structure must fold correctly before L11 can recognize RNA. One argument is that L11 binding is strengthened by the same ionic conditions that stabilize the RNA tertiary structure. Thus L11 binding to the RNA is very weak if Na^+ is the only monovalent ion present, but becomes much stronger if K^+ or NH_4^+ is substituted (Xing and Draper, 1995). This observation provided a way to estimate the NH_4^+-RNA binding affinity: the L11-RNA binding constant was measured at constant total monovalent ion concentration, but with varying ratios of Na^+ and NH_4^+ (Fig. 5a). If we presume that a single NH_4^+ must bind the RNA before L11 can bind, that is,

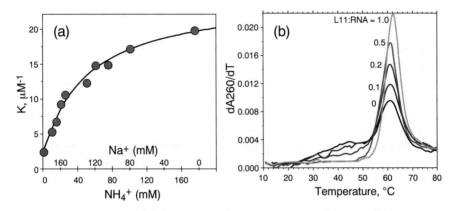

Fig. 5 L11 requires a folded RNA tertiary structure; (a) L11–RNA affinity as a function of NH_4Cl concentration. The curve is calculated assuming a single NH_4^+ ion binds to the RNA with an affinity of $19\,M^{-1}$ before L11 may bind (note that the total monovalent ion concentration has been held constant by varying NaCl concentration). Data are taken from Xing and Draper (1995); (b) Melting profiles of the wild type *E. coli* rRNA domain (Fig. 2a) in the presence of increasing concentrations of protein (RNA was $0.8\,\mu M$). Experimental conditions are given in Xing and Draper (1995).

$$NH_4^+ + RNA \xrightleftharpoons{K_{NH_4^+}} NH_4^+ \bullet RNA$$

$$NH_4^+ \bullet RNA + L11 \xrightleftharpoons{K_{L11}} NH_4^+ \bullet RNA \bullet L11$$

then $K_{NH_4^+}$ is $\sim 20\,M^{-1}$. While this may seem a weak interaction, crown ethers bind K^+ in aqueous solution with similar affinities (Frensdorff, 1971). The stability of the L11–RNA complex is also strongly dependent on Mg^{2+} concentration; one or two ions with apparent affinities of $\sim 10^3\,M^{-1}$ must first bind the RNA to obtain an optimum L11 binding site.

A second argument that RNA tertiary structure promotes L11 recognition comes from melting experiments. The *B. stearothermophilus* L11 used for NMR studies has a melting temperature of $\sim 70°C$, so its effect on RNA melting can be observed without complications from protein denaturation (Xing and Draper, 1995). An example set of melting profiles is shown in Fig. 5b. There is a dramatic stabilization of the RNA structure in the presence of L11, and no RNA unfolding takes place until L11 dissociates from the RNA. This does not mean that L11 contacts the whole RNA, only that L11 directly or indirectly reinforces the tertiary structure.

Homeodomain proteins place helix III in the major groove of DNA. Although L11 uses a similar structure to recognize RNA, it cannot bind in the major groove of an A-form RNA helix, which is much deeper and narrower than a B-form DNA major groove, and in any case the rRNA domain recognizing L11 has only very short stretches of standard Watson-Crick base pairs. Recent NMR studies of an α-helical peptide binding to an RNA hairpin show that mismatches and bulges can open the major groove sufficiently to allow helix access (Battiste *et al.*, 1996). Presumably an RNA minor groove, which is normally too shallow to provide many contacts with an α-helix, could undergo suitable distortions as well. Hydroxyl radical protection studies show that L11 reduces RNA backbone reactivity on either side of the U1060–U1078 mismatch (Rosendahl and Douthwaite, 1993). U1061 is most likely bulged from the helix, as it is reactive towards a carbodiimide reagent (Egebjerg *et al.*, 1990), and A1077 has been implicated in tertiary structure formation (Lu and Draper, 1994). It is therefore probable that L11-C76 helix III is lying in the groove of a distorted helix. The displacement of protections towards the 3′ ends of the helix suggest interactions are taking place in the minor groove (Van Dyke *et al.*, 1982).

5 Targeting of thiazole antibiotics to the L11–RNA complex

Thiostrepton binds the *E. coli* rRNA fragment with an affinity of $\sim 10^6\,\text{M}^{-1}$ (Ryan and Draper, 1989). It also requires an intact tertiary structure for optimum binding affinity, by the same criteria cited above for L11: the affinity is dependent on both monovalent ion type (NH_4^+ and K^+ being more effective than Na^+) and on the presence of Mg^{2+}, and thiostrepton stabilizes the lowest temperature unfolding transition (Wang *et al.*, 1993; Draper *et al.*, 1995).

It has been estimated from equilibrium dialysis experiments that thiostrepton binds about 10^3 fold more tightly to ribosomes containing L11 than to L11-deficient ribosomes (Cundliffe, 1986). This cooperativity can be observed in melting experiments with the rRNA fragment: the ternary thiostrepton–L11–RNA complex melts at higher temperature than either L11–RNA or thiostrepton–RNA complexes (Fig. 6a) (Xing and Draper, 1996). The difference in T_ms of the complexes suggests a cooperativity factor of ~ 200 (evaluated at $\sim 65°C$). Even at these high temperatures, the complex is exchanging slowly on the time scale of the melting experiment (as slow as $0.25\,\text{deg\,min}^{-1}$), suggesting a very tight complex.

Cooperativity between L11 and thiostrepton could arise in two different ways. L11 might trap a specific RNA tertiary structure that is particularly favoured by thiostrepton, and thus indirectly enhance thiostrepton binding affinity. Alternatively, a direct interaction between L11 and thiostrepton bound to the RNA could provide the extra interaction free energy. It is usually difficult to distinguish direct and indirect contributions to the cooperativity between two ligands binding a macromolecule, but in this case we have good evidence for an indirect mechanism. L11-C76 binds rRNA with exactly the same properties as intact L11. Thus we would expect any indirect cooperativity due to protein effects on the RNA conformation to persist with the protein fragment. L11-C76, however, shows no cooperativity with thiostrepton in our melting assay (Fig. 6b) (Xing and Draper, 1996), suggesting that the N-terminal domain of L11 interacts directly with thiostrepton.

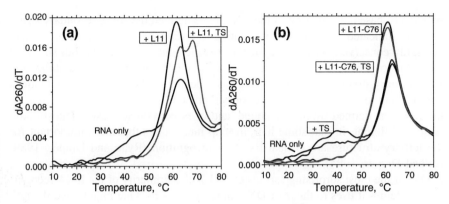

Fig. 6 L11–thiostrepton cooperativity detected in melting experiments. (a) wild type *E. coli* rRNA fragment melted by itself (black), with 0.5 equivalents of L11 (blue), or with 0.5 equivalents of L11 and 4 μM thiostrepton (red). (b) the same experiment repeated with L11-C76 (an additional melting profile with only thiostrepton added (green) is also shown). Data are taken from Xing and Draper (1995, 1996).

Mutations conferring resistance to thiostrepton occur in the two hairpin loops of the L11-binding RNA domain, at positions A1067 and A1095, and thiostrepton protects nucleotides in each loop from reaction with chemical reagents (Rosendahl and Douthwaite, 1994). It is possible to build a model in which stacking of base pairs at the junction of helices brings the two hairpin loops in proximity (Egebjerg *et al.*, 1990). In this model, thiostrepton could slip in between the two loops. Although quite speculative, this model provides a reasonable explanation of data currently available.

Ribosome sensitivity to thiostrepton is not conserved (Amils *et al.*, 1990). Thiostrepton does bind yeast rRNA about 10-fold more weakly than *E. coli* rRNA, but substitution of the yeast sequence for 1055–1104 in *E. coli* ribosomes yields ribosomes that are sensitive to thiostrepton (Thompson *et al.*, 1993). This suggests that differences in the L11 N-terminal domain between eukaryotes and bacteria are an important determinant of thiostrepton sensitivity. Only a single seven-residue sequence of the N-terminal region is unambiguously conserved in all phylogenetic domains, so it is quite plausible that thiostrepton takes advantage of a prokaryote-specific site on L11.

6 Conclusions

Running through these studies of the L11–rRNA interaction are a couple of themes applicable to other RNA–protein interactions from the ribosome. First, the phylogenetic data base of rRNA and protein sequences has been a very useful tool. Almost all the highly conserved protein residues have proved to be either important for proper folding (that is, part of the hydrophobic core or, in the case of gly 71, needed for a sharp turn) or contributors to RNA contacts. Likewise, patterns of base conservation in the RNA have defined the RNA secondary structure and indicated nucleotides contributing to tertiary structure. Some bases may be highly conserved because they make specific contacts with L11, but so far all mutations of conserved nucleotides that weaken L11 binding also destabilize the RNA tertiary structure. L11 may make base specific contact only with nucleotides contributing to the tertiary structure, or perhaps L11 recognizes only the shape of the RNA, as defined by the backbone, and does not contact bases. Thiostrepton is quite different, in that some hairpin loop nucleotides that are unimportant for the tertiary structure are critical for drug recognition (Ryan and Draper, 1989; Lu and Draper, 1994).

One of the most surprising aspects of L11 is the similarity of the RNA recognition strategies it uses to those of DNA binding proteins. One might have thought that the unique tertiary structures of RNA would elicit quite different structures in binding proteins, but at least two other ribosomal proteins are structurally similar to DNA-binding proteins: S17, which has a five strand twisted β-sheet, is similar to the 'oligonucleotide/oligosaccharide binding' class of proteins that include bacteriophage fd gene 5 and cold shock proteins (Jaishree *et al.*, 1996), and the C-terminal domain of L7/L12 contains a helix-turn-helix motif as found in prokaryotic repressors and helices II and III of homeodomains (Rice and Steitz, 1989). One can speculate that ribosomal proteins may have been recruited for other purposes during evolution, and that the ribosome should therefore be a repository of widespread nucleic acid binding motifs. Whether or not this is the case, it certainly appears that there are a small number of themes that recur among nucleic acid binding proteins. The variety of RNA tertiary structures may create only a limited repertoire of shapes suitable for protein recognition.

Acknowledgements

I would like to thank my students whose work on L11 and thiostrepton is cited in this review: Patricia Ryan, Ming Lu, Lance Laing, Yanyan Xing and Debraj GuhaThakurta. Their work has been supported by NIH grant GM29048.

References

Amils, R., Ramirez, L., Sanz, J. L., Marin, I., Pisabarro, A. G., Sanchez, E. and Ureña, D. (1990). In *The Ribosome: Structure, Function, and Evolution* (W. E. Hill, A. Dahlberg, R. A. Garrett, P. B. Moore, D. Schlessinger and J. R. Warner, eds), pp. 645–654. American Society for Microbiology, Washington, D.C.
Battiste, J. L., Mao, H., Rao, N. S., Tan, R., Muhandiram, D. R., Kay, L. E., Frankel, A. D. and Williamson, J. R. (1996). *Science* **273**, 1547–1551.
Beauclerk, A. A. D., Hummel, H., Holmes, D. J., Böck, A. and Cundliffe, E. (1985). *Eur. J. Biochem.* **151**, 245–255.
Berg, J. M. (1990). *Ann. Rev. Biophys. and Biophys. Chem.* **19**, 405–422.
Clubb, R. T., Mizuuchi, M., Huth, J. R., Omichinski, J. G., Savilahti, H., Mizuuchi, K., Clore, G. M. and Gronenborn, A. M. (1996). *Proc. Natl. Acad. Sci. USA* **93**, 1146–1150.
Cole, P. E., Yang, S. K. and Crothers, D. M. (1972). *Biochemistry* **11**, 4358–4368.

Cundliffe, E. (1986). In *Structure, Function, and Genetics of Ribosomes* (B. Hardesty and G. Kramer, eds), pp. 586–604. Springer-Verlag, New York.

Draper, D. E., Xing, Y. and Laing, L. (1995). *J. Mol. Biol.* **249**, 231–238.

Egebjerg, J., Douthwaite, S. R., Liljas, A. and Garrett, R. A. (1990). *J. Mol. Biol.* **213**, 275–288.

El-Baradi, T. T. A. L., Regt, V. H. C. F. D., Einerhand, S. W. C., Teixido, J., Planta, R. J., Ballesta, J. P. G. and Raué, H. A. (1987). *J. Mol. Biol.* **195**, 909–917.

Frensdorff, H. K. (1971). *J. Am. Chem. Soc.* **93**, 600–606.

Hirsch, J. A. and Aggarwal, A. K. (1995). *EMBO J.* **14**, 6280–6291.

Hummel, H. and Böck, A. (1987). *Biochimie* **69**, 857–861.

Jaishree, T. N., Ramakrishnan, V. and White, S. W. (1996). *Biochemistry* **35**, 2845–2853.

Kissinger, C. R., Liu, B., Martin-Blanco, E., Kornberg, T. B. and Pabo, C. O. (1990). *Cell* **63**, 579–590.

Klemm, J., Rould, M. A., Aurora, R., Herr, W. and Pabo, C. O. (1994). *Cell* **77**, 21–32.

Kodandapani, R., Pio, F., Ni, C.-Z., Piccialli, G., Klemsz, M., McKercher, S., Maki, R. A. and Ely, K. R. (1996). *Nature* **380**, 456–460.

Laggerbauer, B., Murphy, F. L. and Cech, T. R. (1994). *EMBO J.* **13**, 2669–2676.

Laing, L. G. and Draper, D. E. (1994). *J. Mol. Biol.* **237**, 560–576.

Li, T., Stark, M. R., Johnson, A. D. and Wohlberger, C. (1995). *Science* **270**, 262–269.

Liljas, A. and Garber, M. (1995). *Curr. Op. in Struc. Biol.* **5**, 721–727.

Lu, M. and Draper, D. E. (1994). *J. Mol. Biol.* **244**, 572–585.

Markus, M., Hinck, A., Huang, S., Draper, D. E. and Torchia, D. A. (1997). *Nature Struc. Biol.* **4**, 70–77.

Musters, W., Gonçalves, P. M., Boon, K., Raué, H. A., van Heerikhuizen, H. and Planta, R. J. (1991). *Proc. Natl. Acad. Sci. USA* **88**, 1469–1473.

Olsen, G. J. and Woese, C. R. (1993). *FASEB J.* **7**, 113–123.

Puglisi, J. D., Pütz, J., Florentz, C. and Giegé, R. (1993). *Nucleic Acids Res.* **21**, 41–49.

Ramakrishnan, V. and Gerchman, S. E. (1991). *J. Biol. Chem.* **266**, 880–885.

Ramakrishnan, V. and White, S. W. (1992). *Nature* **358**, 768–771.

Ramakrishnan, V., Finch, J. T., Graziano, V., Lee, P. L. and Sweet, R. M. (1993). *Nature* **362**, 219–223.

Rice, P. A. and Steitz, T. A. (1989). *Nucleic Acids Res.* **17**, 3757–3762.

Rosendahl, G. and Douthwaite, S. (1993). *J. Mol. Biol.* **234**, 1013–1020.

Rosendahl, G. and Douthwaite, S. (1994). *Nucleic Acids Res.* **22**, 357–363.

Ryan, P. C. and Draper, D. E. (1989). *Biochemistry* **28**, 9949–9956.

Spedding, G. and Cundliffe, E. (1984). *Biochem.* **140**, 453–459.

Stark, M. J. R. and Cundliffe, E. (1979). *J. Mol. Biol.* **134**, 767–779.

Stein, A. and Crothers, D. M. (1976). *Biochemistry* **15**, 160–167.

Thompson, J., Cundliffe, E. and Stark, M. (1979). *Biochem.* **98**, 261–265.

Thompson, J., Schmidt, F. and Cundliffe, E. (1982). *J. Biol. Chem.* **257**, 7915–7917.

Thompson, J., Musters, W., Cundliffe, E. and Dahlberg, A. E. (1993). *Eur. J. Biochem.* **12**, 1499–1504.

Treisman, J., Harris, E., Wilson, D. and Desplan, C. (1992). *BioEssays* **14**, 145–150.

Van Dyke, M. W., Hertzberg, R. P. and Dervan, P. B. (1982). *Proc. Natl. Acad. Sci. USA* **79**, 5470–5474.

Wang, Y.-X., Lu, M. and Draper, D. E. (1993). *Biochemistry* **32**, 12279–12282.

Williamson, J. R., Raghuraman, M. K. and Cech, T. R. (1989). **59**, 871–880.

Xing, Y. and Draper, D. E. (1995). *J. Mol. Biol.* **249**, 319–331.

Xing, Y. and Draper, D. E. (1996). *Biochemistry* **35**, 1581–1588.

Xing, Y., GuhaThakurta, D. and Draper, D. E. (1997). *Nature Struc. Biol.* **4**, 24–27.

9

RNase P and Its Substrate*

LEIF A. KIRSEBOM

Department of Microbiology, Box 581, Biomedical Center, S-751 23 Uppsala, Sweden

Abstract

The ubiquitous RNase P is an endoribonuclease responsible for generating tRNA molecules with matured 5′ termini. Bacterial RNase P consists of an RNA subunit and a protein subunit. The RNA alone cleaves various substrates correctly *in vitro*. The reaction requires the presence of divalent metal ions where Mg^{2+} promotes cleavage most efficiently. The well-conserved GGU motif in *Escherichia coli* RNase P RNA (M1 RNA) base pairs with the 3′-terminal RCCA sequence of a rRNA precursor (interacting residues underlined). This motif is part of an internal loop, P15. Our recent data suggest that the M1 RNA-RCCA interaction plays a role in aligning the scissile bond in the active site such that efficient cleavage is accomplished.

M1 RNA is cleaved spontaneously in specific positions as a result of the addition of divalent ions and two of these cleavage sites are located within P15. The three-dimensional structure of a 31-mer RNA molecule harbouring the M1 RNA P15-loop was recently solved by NMR-spectroscopy. Cleavage of various derivatives of the 31-mer RNA, carrying modifications at specific positions within P15, with Mg^{2+} as well as with Pb^{2+} showed that this small RNA was cleaved at the same positions as native M1 RNA. These findings suggest that the structure of P15 in the 31-mer RNA is a good approximation of the structure of this region in M1 RNA. This is further supported by chemical and enzymatic structural probing data. Additionally, specific chemical groups in P15 involved in coordinating Mg^{2+} (or Pb^{2+}) have been identified by using this 31-mer RNA. These data are discussed in view of the three-dimensional structure of the 31-mer RNA as well as in view of the model of cleavage by RNase P RNA.

---

*Figure 4 for this chapter appears in the colour plate section between p. 144 and p. 145.

THE MANY FACES OF RNA
ISBN 0-12-233210-5

1 Introduction

The tRNA genes in all organisms studied so far are expressed as precursors. To generate biologically functional tRNA molecules these precursors have to be processed. Several enzymes participate in this process and one is RNase P which is responsible for the maturation of the 5′-termini of tRNAs in all cell types studied to date (Altman *et al*., 1995; Kirsebom, 1995; and references therein). However, there are examples of tRNA genes among the Archea and the Eukarya where the first nucleotide in the coding region coincides with the start of transcription (Gupta, 1984; Lee *et al*., 1987).

The endoribonuclease RNase P derived from prokaryotes, eukaryotes as well as from mitochondria is a ribonucleoprotein complex. The biochemical composition (RNA and/or protein) of the activity of chloroplast RNase P, however, is still uncertain (Wang *et al*., 1988). Bacterial RNase P consists of an RNA subunit of approximately 400 nucleotides and a protein subunit of about 14 kDa. The catalytic activity resides in the RNA which under appropriate buffer conditions cleaves its substrate at the correct position without the protein. Thus, RNase P RNA is a naturally isolated trans-acting ribozyme, that is, a catalytic RNA. Cleavage by RNase P RNA as well as by the holoenzyme requires the presence of divalent metal ions, and among these Mg^{2+} promotes cleavage most efficiently. Comparison of cleavage in the absence and in the presence of the protein shows that cleavage in the RNase P RNA alone reaction requires higher concentration of Mg^{2+} (Guerrier-Takada *et al*., 1983). Indeed, it has been suggested that RNase P is a metalloenzyme (Smith and Pace, 1993). The role of the protein is apparently to facilitate product release and its presence also appears to reduce rebinding of the matured tRNA to the enzyme (Reich *et al*., 1988; Tallsjö and Kirsebom, 1993). Furthermore, there is evidence that addition of the protein to the RNA results in a conformational change of the RNA (Vioque *et al*., 1988; Talbot and Altman, 1994a,b). This article will not discuss the function or the structure of the protein subunit of RNase P, but will instead concentrate on what is known about the RNA subunit and its interaction with its substrates.

In bacteria there is only one RNase P enzyme and it is generally believed that this is also the case in eukaryotes; however, it has been suggested that higher vertebrates express multiple isoforms of the RNA subunit (Li and Williams, 1995). Nevertheless, this enzyme has to interact with a large number of different substrates, both tRNA precursors as well as other substrates. In *Escherichia coli* 79 tRNA genes have been identified (Komine *et al*., 1990 and references therein). These are transcribed: (i) as monomeric or multimeric tRNA precursors (harbouring more than one tRNA), (ii) together with rRNA precursors or (iii) together with mRNAs. In the category of other substrates we find several non-tRNA molecules that are cleaved by RNase P. This category includes certain plant viral RNAs, for example TYMV RNA (Guerrier-Takada *et al*., 1988) and the precursor to 10Sa RNA (Komine *et al*., 1994). Both these RNA molecules possess a tRNA-like structure with a coaxially stacked amino acid acceptor-stem and T-stem and in addition they carry a 3′-terminal CCA motif which plays a role in the RNase P RNA–substrate interaction as will be discussed below. The precursors to 4.5 S RNA, bacteriophage M3 RNA and the phage-derived antisense C4 RNA also belong to the non-tRNA substrate category where the two former are stem loop structures (Bothwell *et al.*, 1976a,b) and the latter partially resembles both a precursor tRNA and the precursor to 4.5 S RNA (Hartmann *et al.*, 1995b). These three precursors also carry a 3′-terminal CC(A/C) motif. A stem loop structure within the context of an mRNA was recently suggested to function as a substrate for RNase P *in vivo* (Alifano *et al.*, 1994). This finding is not unexpected given that precursor molecules harbouring only the amino acid acceptor-stem, the T-stem and the T-loop are still cleaved by RNase P (McClain *et al.*, 1987). Taken together, it is clear that RNase P can recognize and process a large number of different substrates correctly in a growing cell. Correct cleavage is essential since miscleavage could affect other steps in the biosynthesis of these RNA molecules as well as their function. This raises the question of how specificity is accomplished. Comparison of the genes encoding the different RNA precursors reveals no consensus sequence at the RNase P cleavage site that could explain how the enzyme identifies its cleavage site. The factors that are important for cleavage site selection may be of help in the design of novel drugs using RNase P RNA as target. Most of the data discussed herein comes from studies using M1 RNA, that is, RNase P RNA derived from *E. coli*.

2 RNase P RNA–substrate interactions

2.1 The 5′-leader sequence

Different tRNA genes are transcribed with varying length of the 5′-leader and very little is known whether this part of a precursor has a function in RNase P cleavage. It has been suggested that the 5′-leader of the precursor to tRNA^{Tyr}Su3 plays a significant role in the formation of a stable enzyme–substrate complex (Guerrier-Takada and Altman, 1993). Changes at the +1 position of this tRNA result in a reduction in its concentration *in vivo* and a change of a C to a U at the −4 position in the corresponding precursor results in a suppression of this effect (Smith, 1976 and references therein). This suggested that nucleotides in the 5′-leader might affect the RNase P processing. However, it is also possible that this change in the 5′-leader influences enzymes involved in the degradation of a tRNA precursor (Ghysen and Celis, 1974). From *in vivo* and *in vitro* studies using various tRNA precursors it is clear that the identity of the nucleotide at the −2 position affects cleavage site selection (Kirsebom and Svärd, 1993; Meinnel and Blanquet, 1995). Cross-linking data suggest that the nucleotide at the −3 position in the leader of tRNA^{Tyr}Su3 precursor derivatives is in close contact with C92 in M1 RNA (Guerrier-Takada *et al.*, 1989; Kufel and Kirsebom, 1994). The importance of C92 for function is demonstrated by the finding that mutant derivatives lacking this nucleotide result in miscleavage of certain tRNA precursors *in vitro* (Guerrier-Takada *et al*., 1989; Kirsebom and Svärd, 1993; Kirsebom, unpublished results). It is very likely that this miscleavage is due to structural effects since addition of C5 results in suppression of the miscleavage pattern and as discussed above it is known that the addition of C5 affects the structure of M1 RNA (RNase P RNA derived from *E. coli*). In conclusion, the 5'-leader apparently plays a role in cleavage by RNase P and this remains to be investigated in more detail. An interesting question is whether the length of 5'-leaders affects the expression of tRNA in a growing cell.

2.2   The tRNA conformation is recognized by RNase P

Shortly after the discovery of an *E. coli* precursor tRNA, structural probing suggested that the overall three-dimensional structure of the tRNA domain of this precursor is not different compared to the structure of the matured tRNA (Smith, 1976). Substitutions which disrupt the tRNA structure lead to defective RNase P processing. This finding indicated that RNase P recognizes the three-dimensional structure of the tRNA domain (Altman *et al*., 1974) and more recent experiments have confirmed this (see for example Altman *et al*., 1995). Thus, the question is what are the determinants in the tRNA domain that makes a tRNA precursor a substrate for RNase P? There are several determinants in a tRNA precursor relevant for RNase P cleavage and, dependent on the nature of the precursor, different determinants play different roles in cleavage by the enzyme. Thus, it is important to understand how the enzyme identifies its cleavage site.

2.3   The roles of the T-loop, T-stem and amino acid acceptor-stem

A common feature of many but not all tRNAs is the -GTUCR- sequence in the T-loop (Fig. 1; Altman *et al*., 1995 and references therein). Modification-interference and cross-linking studies suggest that this part of a precursor tRNA is in close contact with the enzyme (Kahle *et al*., 1990a; Thurlow *et al*., 1991; Nolan *et al*., 1993). These findings are in keeping with the following observations: (i) base-substitutions in the T-loop affect the rate of cleavage as well as cleavage site selection (Altman *et al*., 1995; Svärd and Kirsebom, 1993) and (ii) *in vitro* selection studies suggest that the T-loop in tRNA precursors is a recognition motif, in particular in the absence of the C5 protein (Liu and Altman, 1994). The importance of the T-stem and the T-loop has been shown by the finding that disruption of the well-conserved G53:C61 base pair in three different wild-type tRNA precursors reduces the rate of cleavage significantly (Kirsebom, unpublished data). The contact between RNase P RNA and the T-loop does not appear to be mediated by canonical Watson-Crick interactions (Altman *et al*., 1995; Kirsebom, 1995 and references therein). Modification-interference experiments indicate that the 2'-OH of the riboses at positions 54 and 55 are important for catalysis and it is therefore possible that these hydroxyls interact with specific groups in the RNase P RNA (Gauer and Krupp, 1993). Those groups in M1 RNA might be located in either the G111–A114 region, which is in close contact with

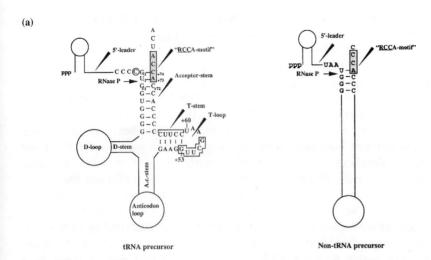

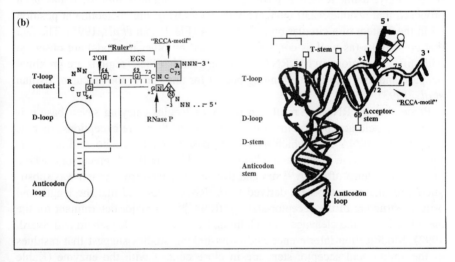

Fig. 1 (a) Schematic illustration of RNase P substrates, a tRNA precursor and a non-tRNA precursor. The RNase P cleavage sites are indicated with arrows. The residue in the tRNA^{Tyr}Su3 precursor which has been UV cross-linked to *E. coli* RNase P RNA is indicated by a circle (for details see the text). The residues in the T-loop which are common to all tRNAs are boxed whereas the shaded nucleotides (the <u>RCC</u>A-motif) in both precursors are those that interact with RNase P RNA. The secondary structures have been taken from the tRNA^{Tyr}Su3 and the 4.5 S RNA precursors. (b) A summary of different determinants important for RNase P cleavage. The 2'OH in the T-stem has been suggested to interact with *Bacillus subtilis* RNase P RNA (Pan *et al.*, 1995). The boxed nucleotides are in close contact with RNase P RNA as revealed by cross-linking (for details see the text). The circled and shaded residues correspond to the same residues as in A. The black residues in the three dimensional representation are part of the 5'-leader.

the well-conserved G53 residue of a tRNA precursor, or in any of the other residues which have been cross-linked to G53 carrying a photoreactive agent (Fig 2; Nolan *et al.*, 1993).

The structure of the T-stem of a tRNA precursor is also suggested to play a significant role in substrate recognition. This is supported by the finding that the introduction of a base pair in the T-stem in a tRNA^{Tyr}Su3 precursor derivative, which is cleaved at two positions, influenced cleavage site selection (Svärd and Kirsebom, 1993). An extension of the T-stem with one base pair also results in a less efficient interaction with the enzyme as assessed by the K_m value (Svärd and Kirsebom, 1992). Modification-interference and cross-linking studies (Kahle *et al.*, 1990a; Thurlow *et al.*, 1991; Nolan *et al.*, 1993) and the *in vitro* selection studies by Liu and Altman (1994) further demonstrate the importance of the T-stem. The cross-linking data suggested that G64 in the T-stem is in close contact with several residues in M1 RNA as demonstrated by main cross-link sites at residues A66, A67, C100, G101, A118 and A248 of M1 RNA. Interestingly, using RNase P RNA derived from *Bacillus subtilis*, it has been proposed that residue A230 interacts with the 2'OH of the nucleotide at position 63 in the T-stem of a circular precursor tRNA (Fig 1; Pan *et al.*, 1995). The fact that cross-links (as well as contacts as detected by other methods) are observed with different regions in M1 RNA might suggest that these regions are proximal in space and that they form a binding pocket for this part of the precursor (Nolan *et al.*, 1993).

Processing by RNase P results in tRNA molecules carrying seven base-pair long amino acid acceptor-stems (Fig. 1). However, there are exceptions, such as tRNA^{His} and tRNA^{SeCys}, which have eight base-pair long amino acid acceptor-stems due to RNase P cleavage. Cleavage of chimeric tRNA precursors, where the acceptor-stems of tRNA^{Tyr}Su3 and tRNA^{Ser} (derived from yeast) was substituted with the acceptor-stem derived from tRNA^{His} showed that the length and primary structure of the acceptor-stem of tRNA^{His} is a major determinant for the identification of the cleavage site (Holm and Krupp, 1992; Kirsebom and Svärd, 1992). Modification-interference and cross-linking studies suggest that residues of the amino acid acceptor-stem are in close contact with the enzyme (Kahle *et al.*, 1990a; Thurlow *et al.*, 1991; Harris *et al.*, 1994). The number of residues in the 5'-half of the acceptor-stem (from the 'base' of the acceptor-stem to the cleavage site, see Fig. 1) appears to play an important role in this process (Svärd and Kirsebom, 1993). The 3'-half of the acceptor-stem is suggested to function as an external guide sequence (EGS; Forster and Altman, 1990). (This latter finding implies that it is in principle possible to target any RNA to function as a substrate for RNase P. For further details see below.) However, part of the EGS in the tRNA^{Tyr}Su3 precursor can be deleted without affecting cleavage at the correct position (Svärd and Kirsebom, 1992). These findings demonstrate that the acceptor-stem is an important determinant for the identification of the cleavage site. Circulary permutated tRNAs harbouring a photoreactive agent at G69 in the acceptor-stem have been cross-linked to several nucleotides in M1 RNA: C10,

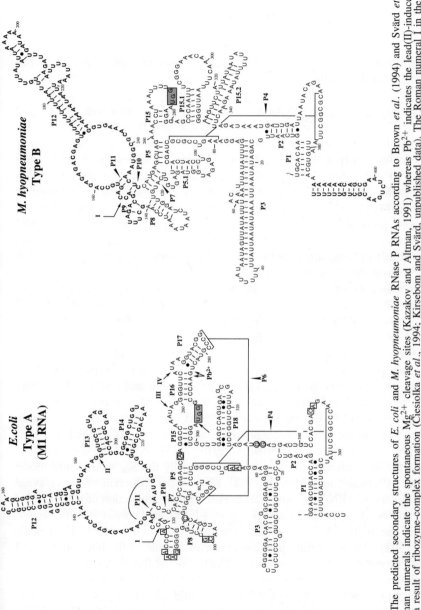

Fig. 2 The predicted secondary structures of *E. coli* and *M. hyopneumoniae* RNase P RNAs according to Brown *et al.* (1994) and Svärd *et al.* (1994). The Roman numerals indicate the spontaneous Mg²⁺ cleavage sites (Kazakov and Altman, 1991) whereas Pb²⁺ indicates the lead(II)-induced cleavage sites as a result of ribozyme-complex formation (Ciesiolka *et al.*, 1994; Kirsebom and Svärd, unpublished data). The Roman numeral I in the case of *M. hyopneumoniae* corresponds to a Pb²⁺ cleavage site (Svärd *et al.*, 1994). Boxed residues correspond to those which have been cross-linked to the T-stem of a tRNA precursor (Nolan *et al.*, 1993) whereas circled residues are those which have been cross-linked to the 5'-leader (Guerrier-Takada *et al.*, 1989; Kufel and Kirsebom, 1994; 1996a,b). The residue A248 has also been cross-linked to the +1 nucleotide of a 5'-modified matured tRNA (Burgin and Pace, 1990). The residues indicated by shaded boxes are involved in Watson–Crick interactions with the 3'-terminal RCCA-motif of a precursor (for details see the text). The type A and type B nomenclature is according to Haas *et al.* (1996). The interaction between C128 and G230 in M1 RNA is in keeping with Mattsson *et al.* (1994).

G64–A67, A192, A248, C252, C253, G332–A334 and U341–A344, suggesting that these play a role in the interaction with the acceptor-stem (Harris *et al*., 1994). From these data it has been suggested that RNase P identifies the site of cleavage by measuring the distance from the T-loop to the cleavage site, which normally is 12 base pairs (with two exceptions, see above; Fig. 1; Kirsebom, 1995 and references therein).

2.4   The role of nucleotides proximal to the cleavage site, in particular the 3′-terminal RCCA motif

The majority of precursor tRNAs derived from *E. coli* harbour both the $-1/+73$ and $-2/+74$ base pairs or just the $-1/+73$ base pair (Fig. 1). Several reports show that these base pairs are not required for correct cleavage. In fact, cleavage occurs preferentially at the expected position even if the cleavage site is single-stranded. It is therefore not base-pairing but rather the identity of the nucleotides at or near the cleavage site which is important for cleavage at the correct position (Kirsebom, 1995 and references therein). The nucleotides at positions -2 to $+1$ have been proposed to constitute one domain and nucleotides at position $+72$ to $+75$ a second domain (Svärd and Kirsebom, 1992). In addition, it has been proposed that the true substrate for RNase P is a precursor with Mg^{2+} bound in the vicinity of the cleavage site (Perreault and Altman, 1993).

More than 80% of all characterized tRNA precursors carry a guanosine at the $+1$ position (Steinberg *et al*., 1993). It has in fact been suggested that a $+1G$ functions as a 'guiding' nucleotide (Kirsebom, 1995 and references therein). The importance of this guanosine is emphasized by the following. A guanosine at the cleavage site of the tRNA[His] precursor is crucial to ensure cleavage only at the correct position. We have suggested that this is due to the presence of the 2-amino group of this guanosine (Tallsjö *et al*., 1996 and references therein). The importance of a 2-amino group, that is, a G at the cleavage site is also evident from work (Kufel and Kirsebom, 1996a) where replacement of a guanosine with an inosine at the cleavage site in a certain context resulted in a shift of the preferred cleavage site. Furthermore, the 2′-hydroxyl of the G+1 is important for the interaction with enzyme as assessed by the K_m value using model substrates (Perreault and Altman, 1992). Finally, the reduction of the cleavage rate of eukaryotic tRNA precursors as a result of the introduction of a 2′ modified nucleotide or of a modified base (m^7G) at $+1$ further demonstrates the significance of the nucleotide 3′ of the cleavage site (Kahle *et al*., 1990b; Kleineidam *et al*., 1993). One intriguing question concerns the role of the nucleotide at $+1$ in those precursors which do not carry a guanosine at this position.

Cross-linking studies using a derivative of the tRNA[Tyr]Su3 precursor, which is cleaved at two positions, revealed that the nucleotide at the -1 position is in close contact with several nucleotides in M1 RNA, in particular A248, A249, C252, C253, A254 as well as G332 and A333. However, the efficiency of cross-linking to these two latter residues was dependent on the positioning of the photoreactive agent relative to the two cleavage sites (Kufel and Kirsebom, 1996a). These findings are in keeping with similar studies (Harris *et al*., 1994; Oh and Pace, 1994). At present it is not known whether there is a direct interaction between the -1 nucleotide and any of these residues in M1 RNA. Interestingly, there is a preference for a pyrimidine, in particular uridine ($\sim$67%), at the -1 position in tRNA precursors derived from *E. coli* (Komine *et al*., 1990).

In *E. coli* all tRNA genes encode the 3'-terminal CCA sequence common to all tRNA whereas there are examples in other bacteria where the CCA is not a part of the gene. This motif has been demonstrated to play a significant role in cleavage by RNase P. Early *in vivo* and *in vitro* experiments showed the importance of the CCA in RNase P cleavage (Kirsebom, 1995 and references therein). In keeping with these findings it was shown that an analogue of the 3'-termini of aminoacyl-tRNAs, puromycin, inhibits the RNase P reaction (Vioque, 1989). Recently, genetic and biochemical experiments demonstrated that the CCA motif interacts with an internal loop in M1 RNA (Kirsebom and Svärd, 1994; LaGrandeur *et al*., 1994; Oh and Pace, 1994; Svärd *et al*., 1996). This interaction is mediated through base-pairing between the two Cs of the CCA and G292 and G293, two well-conserved residues among bacterial RNase P RNAs (Kirsebom and Svärd, 1994; Svärd *et al*., 1996). The data of Tallsjö *et al*. (1996) suggests that this interaction also includes the discriminator base (the residue at position +73 in a tRNA precursor; Fig. 1) in the precursor and the well-conserved U294 in M1 RNA. This interaction has been designated the '<u>R</u>CCA-M1 RNA' interaction (interacting residues underlined, see Fig. 5; Kufel and Kirsebom, 1996a). Cross-linking and modification interference experiments using various RNase P RNA species as well as a phylogenetic comparative mutational analysis suggest that this interaction is not only relevant to *E. coli* RNase P RNA but also to other bacterial RNase P RNAs (LaGrandeur *et al*., 1994; Oh and Pace, 1994; Svärd *et al*., 1996). However, for RNase P RNA derived from *Mycoplasma hyopneumoniae* the situation is less clear. Here, it appears that an interaction between <u>C</u>74 in the 3'-terminal R<u>C</u>CA motif (interacting residue underlined) and RNase P RNA is established in a similar way as discussed above, at least in the case of using an *E. coli* tRNA^{Ser} precursor, while the interaction with the second C at position 75 in the precursor is less clear (Svärd *et al*., 1996). It should also be emphasized that there might be some other species variations; for example the residue which corresponds to U294 in M1 RNA is replaced by an adenosine in the RNase P RNA derived from a few bacterial species such as *Thermus thermophilus* (Hartmann and Erdmann, 1991). Also, it is interesting to note that RNase P RNA derived from *Chlamydia* spp. and cyanobacteria show significant structural differences in the region that interacts with the 3'-terminal R<u>CC</u>A motif of the substrate (Fig. 3; Vioque, 1992; Herrmann *et al*., 1996 and Törner *et al*. unpublished results). This might be related to the fact that the tRNA genes from these species characterized so far do not encode the 3'-terminal CCA motif (Cousineau *et al*., 1992; Steinberg *et al*., 1993).

136   **L. A. Kirsebom**

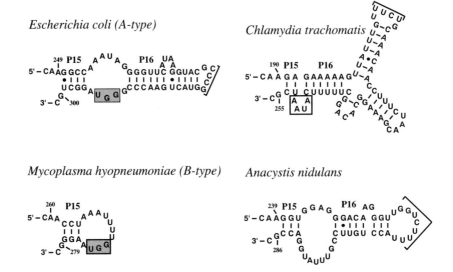

Fig. 3 Suggested secondary structures of the P15 to P17 regions in *E. coli*, *M. hyopneumoniae*, *C. trachomatis* (Herrmann *et al.*, 1996) and *A. nidulans* (Banta *et al.*, 1992). Shaded regions in the two former denote nucleotides interacting with the 3′-terminal RCCA-motif of a precursor (for details see the text) whereas the brackets indicate residues involved in the P6 interaction (see Fig. 2).

The 'RCCA-M1 RNA' interaction plays a significant role both in the identification of the cleavage site and in the kinetics of cleavage by influencing product release (Kirsebom, 1995; Tallsjö *et al.*, 1996). The importance of this interaction is further emphasized by the finding that a tRNA precursor lacking the 3′-terminal CCA motif is cross-linked to different residues in M1 RNA compared with when this sequence element is present (Guerrier-Takada *et al.*, 1989). This is in keeping with the suggestion that the tRNA domain of a precursor might bind to M1 RNA in different modes (Westhof and Altman, 1994; Westhof *et al.*, 1996). In this context, it has also been suggested that the active conformation of M1 RNA is dependent on the identity of the substrate (Guerrier-Takada *et al.*, 1989; Kirsebom and Svärd, 1993). We have suggested that different cleavage sites are aligned differently in the active site, possibly as a result of the different binding modes of a substrate to M1 RNA. In this model the 'RCCA-M1 RNA' interaction plays a significant role, perhaps by initiating docking of the substrate into the active site (Kufel and Kirsebom, 1996a,b).

2.5 Non-tRNA substrates

The non-tRNA substrates cleaved by RNase P, such as the precursor to 4.5 S RNA and artificial substrates, consist of a stem structure which is thought to mimic the coaxial stacked acceptor-stem stacked on the T-stem (Fig. 1; McClain *et al.*, 1987; Forster and Altman, 1990; Peck-Miller and Altman, 1991). In the absence of the C5 protein these substrates interact less efficiently with the enzyme compared to tRNA precursors as determined by the K_m values. This might suggest that the D-stem, D-loop, anticodon stem and anticodon loop help to anchor the tRNA precursor on the enzyme. By using mutant M1 RNAs carrying deletions of varying sizes, Guerrier-Takada and Altman (1992) showed that the presence of the upper domain of M1 RNA (nucleotides 94 to 204, Fig. 2) is required for cleavage of a tRNA precursor but not for cleavage of 4.5 S RNA. These authors proposed that this part of M1 RNA carries a region that participates in an unfolding of the tertiary structure of the tRNA^{Tyr}Su3 precursor. (In this context it is interesting to note that the holoenzyme is more versatile than the RNA enzyme (Liu and Altman, 1994).) These observations, together with other data, show that cleavage by RNase P is dependent on the identity of the substrate and might suggest that tRNA precursors and non-tRNA substrates bind to M1 RNA differently as has been discussed elsewhere (Guerrier-Takada et al., 1989; Kirsebom and Svärd, 1993). This is supported by a recent report suggesting that B. subtilis RNase P RNA binds a non-tRNA substrate differently compared to a tRNA precursor (Pan and Jakacka, 1996). The cleavage of non-tRNA substrates is clearly in keeping with the findings that the amino acid acceptor-stem and the T-stem of a tRNA precursor are important determinants in cleavage by RNase P.

3 RNase P RNA structure and function

3.1 Differences in cleavage by different RNase P RNAs as a function of structure

The RNA component is the catalytic subunit of bacterial RNase P and it is generally believed that this is also the case for RNase P derived from other sources. Knowledge about the structure of RNase P RNA is of importance in order to understand its function. Today over 100 sequences from all of the 11 major phylogenetic branches are available (Haas et al., 1996). Although RNase P RNAs derived from various species differ significantly in their primary structures it is possible to fold these into similar secondary structure models (Fig. 2). Examination of these models reveals that the structure of the M1 RNA type (A-type, RNase P RNA derived from Gram-negative and high G-C content Gram-positive bacteria) is different compared to the structure of the RNase P RNA derived from low G-C content Gram-positive bacteria as exemplified by *Mycoplasma hyopneumoniae* [B-type (Haas et al., 1996; Svärd et al., 1994; Mattsson et al., unpublished data)]. The most apparent differences are: (i) the absence of the P6 interaction in the B-type; instead the B-type harbours P5.1, P15.1 and P15.2 in the core; (ii) RNase P RNA of the B-type also lacks P13 and P14 whereas P10.1 as well as P19 are absent in the A-type (the two latter not shown in the figure since *M. hyopneumoniae* lacks these helicies); (iii) A-type RNase P RNA carries an internal loop between P15 and P16 whereas in the B-type this element is represented by a loop structure. As discussed above, irrespective of A- or B-type, this domain of RNase P RNA plays an important role in the interaction with the 3'-terminal RCCA end of the substrate. Although the secondary structures of these two classes of RNase P RNA differ it has been suggested that their tertiary structures are similar (Haas *et al.*, 1991). However, *M. hyopneumoniae* RNase P RNA (B-type) miscleaved an *E. coli* tRNA precursor whereas M1 RNA and *Mycobacterium tuberculosis* RNase P RNA (both of the A-type) cleaved this precursor correctly (Svärd *et al.*, 1996; and unpublished data). Additionally, by comparing cleavage by M1 RNA and *M. hyopneumoniae* RNase P RNA under identical buffer conditions a significant difference in K_m is revealed (Svärd *et al.*, 1994). This might reflect differences in the structures of these two RNase P RNAs that result in a less efficient interaction of the *M. hyopneumoniae* derived RNase P RNA with the substrate. As discussed, it is plausible that the 3'-terminal RCCA motif of a precursor interacts slightly differently with a B-type RNase P RNA such as *M. hyopneumoniae* RNase P RNA compared to the interaction with M1 RNA (Svärd *et al.*, 1996). Thus it can be seen that phylogenetic analysis is an important tool for investigating the structure function relationship of RNase P RNA. However, for a more detailed understanding of the function in relation to its structure other tools are required. For instance, compensatory mutation analysis has been shown to be a powerful method for the investigation of interactions within and between RNA molecules (see above and Gesteland and Atkins, 1993; Tallsjö *et al.*, 1993; Mattsson *et al.*, 1994).

3.2 The 'RCCA-M1 RNA' interaction and divalent metal ion binding

Crystallographic structures of RNase P RNA alone or in complex with its substrate are not yet available. However, two working models of the three-dimensional structure of M1 RNA in complex with a precursor tRNA have been presented (Harris *et al.*, 1994; Westhof and Altman, 1994; Westhof *et al.*, 1996). Both these models take into account data from chemical and enzymatic protection experiments, intra- and intermolecular cross-linking studies, phylogenetic analysis, studies of the activities of mutants and the kinetics of reactions catalysed as a result of ribozyme-substrate (RS) complex formation. For a detailed discussion of these models see the original papers as well and the work of Kirsebom (1995) and Hartmann and his colleagues (Hardt *et al.*, 1995; 1996). In the model proposed by Harris *et al.* (1994) the amino acid acceptor-stem remains folded in the complex whereas the opposite is the case in the other model, which takes into account the data suggesting Watson-Crick base-pairing between the 3'-terminal 'RCCA-motif' of the precursor and the internal loop located between helices P15 and P16, here defined as the P15-loop (see above). The following discussion will focus on the structure and function of the P15-loop. I would like also to emphasize that given that the major part of the amino acid acceptor-stem remains folded and the only unfolded base pairs in the RS-complex would be $-2/+74$ and/or $-1/+73$, then the GGU-motif of the P15-loop would be part of (or very close to) the active centre of RNase P RNA (Kirsebom and Svärd, 1994).

Magnesium(II), or lead(II), induces spontaneous cleavage at specific positions on RNase P RNA (Kazakov and Altman, 1991; Ciesiolka *et al*., 1994; Zito *et al*., 1993). Two of these cleavage sites, III and V (Fig. 2), are located within the P15-loop. This suggests that Mg^{2+} ions are bound in the vicinity of P15 and it has been discussed in several reports that this Mg^{2+}-ion(s) is important for catalysis (Kazakov and Altman, 1991; Kirsebom and Svärd, 1993; Kufel and Kirsebom, 1996b). Substitutions within P15, as well as formation of the RS-complex, result in a change in the lead(II)-induced cleavage pattern (Ciesiolka *et al*., 1994; Kufel and Kirsebom, 1996b; Kirsebom and Svärd, unpublished data). These findings suggest that the structural integrity of the P15-loop is important for divalent metal ion binding and that a conformational change(s) of this internal loop is induced as a consequence of RS-complex formation. In fact, formation of a RNase P RNA-tRNA precursor complex gives a novel Pb^{2+}-induced cleavage in the P16 helix (Fig. 2; Ciesiolka *et al*., 1994; Kirsebom and Svärd, unpublished data). From this it is clear that knowledge of the three-dimensional structure of P15 as definition of the Mg^{2+}-binding site(s) is of importance.

A model RNA harbouring the P15-loop has been studied with respect to its structure as well as divalent metal ion binding (Kufel, 1996). The structure of this 31-mer RNA was determined by NMR-spectroscopy (Fig. 4, Glemarec *et al*., 1996). We probed the structure of various derivatives of this model RNA (enzymatically, chemically as well as by Mg^{2+} and Pb^{2+} cleavage) and compared these results with the data obtained using full-size M1 RNA. From the combined data we can conclude that the P15-loop in the model RNA is folded in approximately the same way as in intact M1 RNA, forming an autonomous divalent metal ion-binding domain. Further, the residues involved in base-pairing with the substrate are indeed accessible both in the model RNA as well as in full-size RNase P RNA, as revealed by hybridization of a complementary deoxyoligoribonucleotide under native conditions and RNase H cleavage. By studying the effect on Mg^{2+} and Pb^{2+}-induced cleavage as a result of incorporating modified nucleotides (Rp-αS- and deoxy-substituted) at specific positions in the model RNA it has been possible to position two divalent metal ions close to U8 and G25. These residues correspond to U257 and G293, respectively, in M1 RNA (see Figs 2 and 5).

A possibility is that a Mg^{2+}-ion(s) might be involved in stabilizing the 'RCCA-M1 RNA' (Kufel and Kirsebom, 1994). This could be accomplished in such a way that a Mg^{2+}-ion(s) bound in the P15-loop stabilizes a conformation required for an efficient interaction with the substrate. Alternatively, it may be that the 'RCCA-M1 RNA' interaction results in a change in conformation of the P15-loop. Thus, as a consequence a re-coordination of a Mg^{2+}-ion(s) would result such that correct cleavage occurs. This is consistent with the fact that a novel Pb^{2+} cleavage site in RNase P RNA has been observed as a result of RS-complex formation (discussed above). These two alternatives are not mutually exclusive and are in agreement with the findings that Mg^{2+} plays an important role in the structural organization of the P15-loop (Westhof *et al*., 1996) as well as that the

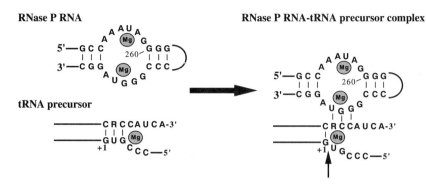

Fig. 5 Illustration of the base-pairing between RNase P RNA and its substrate. The R in the precursor structure denotes a G or an A indicating a G–U wobble base pair or an A–U Watson-Crick base pair in the RS-complex. There are three Mg^{2+}-ions depicted in the figure; two of which have been identified in the P15-loop (Kufel, 1996) and one which is bound in the vicinity of the cleavage site on the precursor (Kazakov and Altman, 1991; Perreault and Altman, 1993). This does not imply that these three Mg^{2+}-ions are involved in the chemistry of cleavage but that they might be. Based on the data of Warnecke *et al.* (1996) there are at present no reasons to invoke more than two Mg^{2+}-ions in the actual cleavage reaction. The secondary structures have been taken from the tRNA^{Tyr}Su3 precursor and M1 RNA.

structural integrity of this internal loop is of importance for Mg^{2+} binding (Kufel and Kirsebom, 1996b). For a discussion of other potentially important Mg^{2+}-ions the reader may refer to Harris and Pace (1995), Hardt *et al.* (1995) and Beebe *et al.* (1996).

4 Summary

Our present view of how *E. coli* RNase P identifies its cleavage site is the following: RNase P recognizes the conformation of the tRNA domain of a tRNA precursor. The enzyme makes contact with nucleotides in the T-loop as well as with the amino acid acceptor-stem and T-stem. Here a stem-loop structure mimicking the amino acid acceptor-stem is sufficient; however, this gives a less efficient interaction as assessed by comparison of K_m values (McClain *et al.*, 1987; Perreault and Altman, 1992). The 3′-terminal A(/G)CC of the precursor establishes a Watson-Crick interaction with G292, G293 and U294 in M1 RNA (when the nucleotide at +73 is a G it may still form a wobble base pair with U294). This is suggested to result in an unfolding of the precursor thereby exposing the cleavage site. In addition, this interaction is suggested to result in a re-coordination of a Mg^{2+}-ion(s) such that cleavage is accomplished at the correct position (see Fig. 5). Here it is conceivable that the identity of nucleotides at or near the cleavage site are important for Mg^{2+} binding. Correct coordination of Mg^{2+} is in keeping with the finding that a tRNA precursor carrying a single Sp-diastereomer at the cleavage site resulted in miscleavage (Warnecke *et al.*, 1996). Interestingly, it has been suggested that the true substrate for RNase P is a precursor with a Mg^{2+} bound in the vicinity of the cleavage site (Perreault and Altman, 1993). The different determinants for cleavage site selection are summarized in Fig. 1. With respect to different models for the mechanism of cleavage, discussions can be found in Smith and Pace (1993), Westhof and Altman (1994) as well as in Warnecke *et al*. (1996).

5 Concluding remarks

Gene inactivation by using ribozymes has opened new perspectives for the treatment of various diseases (for a review see Hartmann *et al*., 1995a). It has been demonstrated by using the external guide sequence concept (EGS; Forster and Altman, 1990) that it is possible to target and subsequently cleave specific mRNAs by RNase P in a growing bacteria as well as in cell cultures (Guerrier-Takada *et al*., 1995; Liu and Altman, 1995; Li and Altman, 1996). Furthermore, since RNase P is an essential enzyme it is also a potential target to combat pathogens. Thus RNase P RNA has a potential usage both in relation to diagnostics as well as a target for novel inhibitors. In relation to diagnostics it has been demonstrated that RNase P RNA genes derived from closely related bacterial species show larger sequence differences than the genes encoding 16 S rRNA (Svärd *et al*., 1994; Herrmann *et al*., 1996; Törner *et al*., unpublished data). While there are several antibiotics that inhibit the function of the ribosome through interaction with the rRNA there is only one antibiotic identified so far that inhibits cleavage of a precursor tRNA by RNase P RNA *in vitro* (Vioque, 1989). Thus, understanding the function and structure of RNase P is clearly of importance to further develop the use of RNase P RNA in gene inactivation as well as to develop diagnostic protocols and identify (and/or design) new drugs directed against RNase P.

Acknowledgements

I am grateful to all the people I have had opportunity to work with over the years. Drs E. Bridge, S. Dasgupta and J. Kufel are acknowledged for critical reading of the manuscript. The ongoing work in my laboratory is supported by grants from the Swedish Natural Research Council and the Swedish Research Council for Engineering Sciences.

References

Alifano, P., Rivellini, F., Piscitelli, C., Arraiano, C. M., Bruni, C. B. and Carlomagno, M. S. (1994). *Genes Dev.* **8**, 3021–3031.
Altman, S., Bothwell, A. L. M. and Stark, B. C. (1974). *Brookhaven Symposia in Biology* **26**, 12–25.
Altman, S., Kirsebom, L. A. and Talbot, S. (1995). In *tRNA: Structure, Biosynthesis, and Function* (D. Söll and U. RajBhandary, eds), pp 67–78. American Society for Microbiology, Washington DC.

Banta, A. B., Haas, E. S., Brown, J. W. and Pace, N. R. (1992). *Nucleic Acids Res.* **20**, 911.
Beebe, J. A., Kurz, J. C. and Fierke, C. A. (1996). *Biochemistry* **35**, 10493–10505.
Bothwell, A. L. M., Garber, R. L. and Altman, S. (1976a). *J. Biol. Chem.* **251**, 7709–7716.
Bothwell, A. L. M., Stark, B. C. and Altman, S. (1976b). *Proc. Natl. Acad. Sci. USA* **73**, 1912–1916.
Brown, J. W., Haas, E. S., Gilbert, D. G. and Pace, N. R. (1994). *Nucleic Acids Res.* **22**, 3660–3662.
Burgin, A. B. and Pace, N. R. (1990). *EMBO J.* **9**, 4111–4118.
Ciesiolka, J., Hardt, W-D., Schlegl, J., Erdmann, V. A. and Hartmann, R. K. (1994). *Eur. J. Biochem.* **219**, 49–56.
Cousineau, B., Cerpa, C., Lefebvre, J. and Cedergren, R. (1992). *Gene* **120**, 33–41.
Forster, A. C. and Altman, S. (1990). *Science* **249**, 783–786.
Gaur, R. K. and Krupp, G. (1993). *Nucleic Acids Res.* **21**, 21–26.
Gesteland, R. A. and Atkins, J. A., eds (1993). *The RNA World.* New York: Cold Spring Harbor Laboratory Press.
Ghysen, A. and Celis, J. E. (1974). *J. Mol. Biol.* **83**, 333–351.
Glemarec, C., Kufel, J., Földesi, A., Maltseva, T., Sandström, A., Kirsebom, L. A. and Chattopadhyaya, J. (1996). *Nucleic Acids Res.* **24**, 2022–2035.
Guerrier-Takada, C., Gardiner, K., Marsh, T., Pace, N. and Altman, S. (1983). *Cell* **35**, 849–857.
Guerrier-Takada, C., van Belkum, A., Pleij, C. W. A. and Altman, S. (1988). *Cell* **53**, 267–272.
Guerrier-Takada, C., Lumelsky, N. and Altman, S. (1989). *Science* **286**, 1578–1584.
Guerrier-Takada, C. and Altman, S. (1992). *Proc. Natl. Acad. Sci. USA* **89**, 1266–1270.
Guerrier-Takada, C. and Altman, S. (1993). *Biochemistry* **32**, 7152–7161.
Guerrier-Takada, C., Li, Y. and Altman, S. (1995). *Proc. Natl. Acad. Sci. USA* **92**, 11115–11119.
Gupta, R. (1984). *J. Biol. Chem.* **259**, 9461–9471.
Haas, E. S., Morse, D. P., Brown, J. W., Schmidt, F. J. and Pace, N. R. (1991). *Science* **254**, 853–856.
Haas, E. S., Brown, J. W., Pitulle, C. and Pace, N. R. (1994). *Proc. Natl. Acad. Sci. USA* **91**, 2527–2531.
Haas, E. S., Banta, A. B., Harris, J. K., Pace, N. R. and Brown, J. W. (1996). *Nucleic Acids Res.* **24**, 4775–4782.
Hardt, W-D., Warnecke, J. M., Erdmann, V. A. and Hartmann, R. K. (1995). *EMBO J.* **14**, 2935–2944.
Hardt, W-D., Erdmann, V. and Hartmann, R. K. (1996). *RNA* **2**, 1189–1198.
Harris, M. E., Nolan, J. M., Malhotra, A., Brown, J. W., Harvey, S. C. and Pace, N. R. (1994). *EMBO J.* **13**, 3953–3963.
Harris, M. E. and Pace, N. R. (1995). *RNA* **1**, 210–218.
Hartmann, R. K. and Erdmann, V. (1991). *Nucleic Acids Res.* **19**, 5957–5964.
Hartmann, R. K., Krupp, G. and Hardt, W-D. (1995a). *Biotech. Ann. Rev.* **1**, 215–265.
Hartmann, R. K., Heinrich, J., Schlegl, J. and Schuster, H. (1995b). *Proc. Natl. Acad. Sci. USA* **92**, 5822–5826.
Herrmann, B., Winqvist, O., Mattsson, J. G. and Kirsebom, L. A. (1996). *J. Clin. Microbiol.* **34**, 1897–1902.
Holm, P. S. and Krupp, G. (1992). *Nucleic Acids Res.* **20**, 421–423.
Kahle, D., Wehmeyer, U. and Krupp, G. (1990a). *EMBO J.* **9**, 1929–1937.
Kahle, D., Wehmeyer, U. and Krupp, G. (1990b). *Nucleic Acids Res.* **18**, 837–843.
Kazakov, S. and Altman, S. (1991). *Proc. Natl. Acad. Sci. USA* **88**, 9193–9197.
Kirsebom, L. A. (1995). *Mol. Microbiol.* **17**, 411–420.

Kirsebom, L. A. and Altman, S. (1989). *J. Mol. Biol.* **207**, 837-840.
Kirsebom, L. A. and Svärd, S. G. (1992). *Nucleic Acids Res.* **20**, 425-432.
Kirsebom, L. A. and Svärd, S. G. (1993). *J. Mol. Biol.* **231**, 594-604.
Kirsebom, L. A. and Svärd, S. G. (1994). *EMBO J.* **13**, 4870-4876.
Kleineidam, R. G., Pitulle, C., Sproat, B. and Krupp, G. (1993). *Nucleic Acids Res.* **21**, 1097-1101.
Knap, A. K., Wesolowski, D. and Altman, S. (1990). *Biochimie* **72**, 779-790.
Komine, Y., Adachi, T., Inokuchi, H. and Ozeki, H. (1990). *J. Mol. Biol.* **212**, 579-598.
Komine, Y., Kitabatake, M., Yokogawa, T., Nishikawa, K. and Inokuchi, H. (1994). *Proc. Natl. Acad. Sci. USA* **91**, 9223-9227.
Kufel, J. (1996). *Thesis.* Acta Universitatis Upsaliensis, Uppsala.
Kufel, J. and Kirsebom, L. A. (1994). *J. Mol. Biol.* **244**, 511-521.
Kufel, J. and Kirsebom, L. A. (1996a). *Proc. Natl. Acad. Sci. USA* **93**, 6085-6090.
Kufel, J. and Kirsebom, L. A. (1996b). *J. Mol. Biol.* **263**, 685-698.
LaGrandeur, T. E., Hüttenhofer, A., Noller, H. F. and Pace, N. R. (1994). *EMBO J.* **13**, 3945-3952.
Lee, B. J., De La Pena, P., Tobian, J. A., Zasloff, M. and Hatfield, D. (1987). *Proc. Natl. Acad. Sci. USA* **84**, 6384-6388.
Li, K. and Williams, S. (1995). *J. Biol. Chem.* **270**, 25281-25285.
Li, Y. and Altman, S. (1996). *Nucleic Acids Res.* **24**, 835-842.
Liu, F. and Altman, S. (1994). *Cell* **77**, 1093-1100.
Liu, F. and Altman, S. (1995). *Genes Dev.* **9**, 471-480.
Mattsson J. G., Svärd, S. G. and Kirsebom, L. A. (1994). *J. Mol. Biol.* **241**, 1-6.
McClain, W. H., Guerrier-Takada, C. and Altman, S. (1987). *Science* **238**, 527-530.
Meinnel, T. and Blanquet, S. (1995). *J. Biol. Chem.* **270**, 15908-15914.
Nolan, J. M., Burke, D. M. and Pace, N. R. (1993). *Science* **261**, 762-765.
Oh, B-K. and Pace, N. R. (1994). *Nucleic Acids Res.* **22**, 4087-4094.
Pan, T. and Jakacka, M. (1996). *EMBO J.* **15**, 2249-2255.
Pan, T., Loria, A. and Zhong, K. (1995). *Proc. Natl. Acad. Sci. USA* **92**, 12510-12514.
Peck-Miller, K. A. and Altman, S. (1991). *J. Mol. Biol.* **221**, 1-5.
Perreault, J-P. and Altman, S. (1992). *J. Mol. Biol.* **226**, 399-409.
Perreault, J-P. and Altman, S. (1993). *J. Mol. Biol.* **230**, 750-756.
Reich, C., Olsen, G. J., Pace, B. and Pace, N. R. (1988). *Science* **239**, 178-181.
Smith, J. D. (1976). *Progress in Nucl. Acids Res. and Mol. Biol.* **16**, 25-73.
Smith, D. and Pace, N. R. (1993). *Biochemistry* **32**, 5273-5281.
Steinberg, S., Misch, A. and Sprinzl, M. (1993). *Nucleic Acids Res.* **21**, 3011-3015.
Svärd, S. and Kirsebom, L. A. (1992). *J. Mol. Biol.* **227**, 1019-1031.
Svärd, S. and Kirsebom, L. A. (1993). *Nucleic Acids Res.* **21**, 427-434.
Svärd, S. G., Mattsson, J. G., Johansson, K-E. and Kirsebom, L. A. (1994). *Mol. Microbiol.* **11**, 849-859.
Svärd, S. G., Kagardt, U. and Kirsebom, L. A. (1996). *RNA* **2**, 463-472.
Talbot, S. J. and Altman, S. (1994a). *Biochemistry* **33**, 1399-1405.
Talbot, S. J. and Altman, S. (1994b). *Biochemistry* **33**, 1406-1411.
Tallsjö, A. and Kirsebom, L. A. (1993). *Nucleic Acids Res.* **21**, 51-57.
Tallsjö, A., Svärd, S. G., Kufel, J. and Kirsebom, L. A. (1993). *Nucleic Acids. Res.* **21**, 3927-3933.
Tallsjö, A., Kufel, J. and Kirsebom, L. A. (1996). *RNA* **2**, 299-307.
Thurlow, D. L., Shilowski, D. and Marsh, T. L. (1991). *Nucleic Acids Res.* **19**, 885-891.
Vioque, A. (1989). *FEBS Lett.* **246**, 137-139.
Vioque, A. (1992). *Nucleic Acids Res.* **20**, 6331-6337.
Vioque, A., Arnez, J. and Altman, S. (1988). *J. Mol. Biol.* **202**, 835-848.
Wang, M. J., Davis, N. W. and Gegenheimer, P. (1988). *EMBO J.* **7**, 1567-1574.

Warnecke, J. M., Fürste, J. P., Hardt, W-D., Erdmann, V. and Hartmann, R. K. (1996). *Proc. Natl. Acad. Sci. USA* **93**, 8924-8928.
Westhof, E. and Altman, S. (1994). *Proc. Natl. Acad. Sci. USA* **91**, 5133-5137.
Westhof, E., Wesolowski, D. and Altman, S. (1996). *J. Mol. Biol.* **258**, 600-613.
Zito, K., Hüttenhofer, A. and Pace, N. R. (1993). *Nucleic Acids Res.* **21**, 5916-5920.

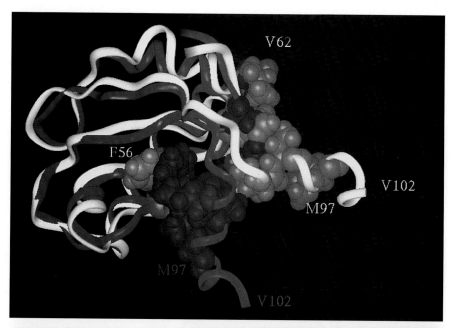

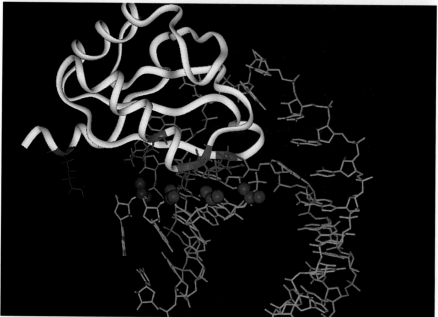

Chapter 5, Fig. 6 (a) Binding of UIA to its RNA targets requires a sharp motion of Helix C centred around residue 90. This motion disrupts a hydrophobic patch on the surface of the β-sheet (red) and creates a second hydrophobic patch involving His10, Leu 41 and Val 62 (silver). (b) Interaction of unique basic side chains within loop 1 and loop 3 of U1A (purple) with the RNA phosphodiester backbone.

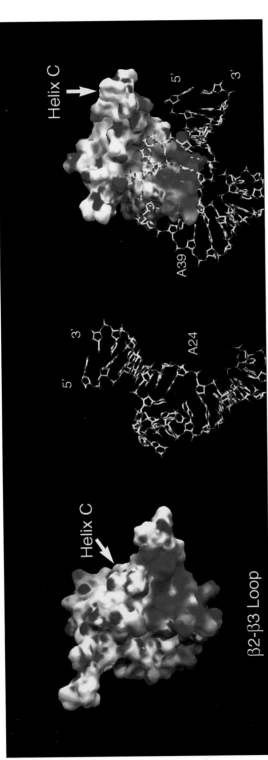

Chapter 5, Fig. 7   Structures of the free U1A protein (right), the free RNA internal loop (centre) and the protein–RNA complex (left) (Allain *et al.*, 1996). Protein binding induces a dramatic change in the RNA conformation and requires a sharp reorientation of Helix C at the carboxy end of U1A.

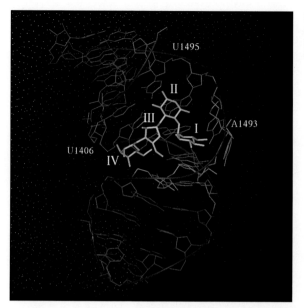

Chapter 7, Fig. 6 Overall structure of the A-site RNA-paromomycin complex, determined from NMR data.

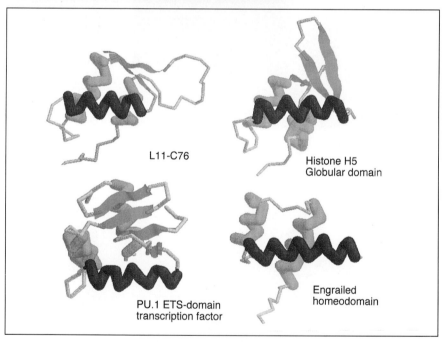

Chapter 8, Fig. 1 The C-terminal fragment of ribosomal protein L11 (L11-C76) and its similarity to homeodomain-like DNA-binding proteins. All four proteins have three α-helices that can be approximately superimposed. Helix III of each protein is coloured red-orange, and oriented with its N-terminus on the left. The other two α-helices are in orange, and β-sheets are in green. L11-C76 structure was determined by NMR (Xing *et al.*, 1997). Engrailed homeodomain and PU.1 ETS domain structures are taken from co-crystals with bound DNA (Kissinger *et al.*, 1990; Kodandapani *et al.*, 1996), and the globular domain of histone H5 is a crystal structure of protein alone (Ramakrishnan *et al.*, 1993).

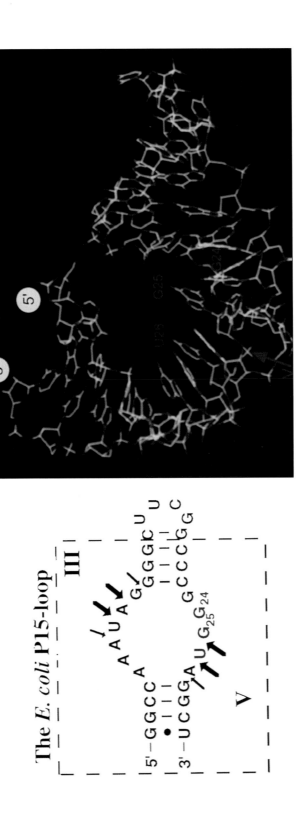

Chapter 9, Fig. 4 (a) A model RNA harbouring the P15-loop of M1 RNA. The structure within the box indicated with dashed lines is identical to the structure of this region in full-size M1 RNA (see Fig. 2). Arows indicate Mg^{2+} and Pb^{2+}-induced cleavage sites whereas the Roman numerals represent the spontaneous Mg^{2+}-induced cleavage sites in full-size M1 RNA (see Fig. 2). The red residues (G24 through U26) denote nucleotides interacting with the 3'-terminal RCCA-motif of a precursor. These correspond to G292, G293 and U294 in full-size M1 RNA. (b) The three dimensional structure of the model RNA carrying the P15-loop (Glemarec *et al.*, 1996). The red-labelled residues are the same as in (a), whereas the red arrows indicate the main Mg^{2+} induced cleavage sites (III 3' of U8 and V 3' of G25 corresponding to U256 and G293 in M1 RNA, respectively) in the P15-loop in full-size M1 RNA. For details see the text.

10

Interfering with Gene Function at the RNA Level with Oligozymes—the Development of New Tools and Therapeutics

BRIAN S. SPROAT[1], SHAJI T. GEORGE[2], MICHAEL MA[2] AND ALLAN R. GOLDBERG[2]

[1] INNOVIR GmbH, Olenhuser Landstrasse 20B, 37124 Rosdorf, Germany; [2] INNOVIR Laboratories Inc., 510 E. 73rd Street, New York, NY 10021, USA

Abstract

The hammerhead ribozyme motif, discovered in certain plant pathogens, is a member of a small class of naturally occurring catalytic RNAs. In nature an intramolecular reaction leads to cis-cleavage; however, it was shown that the naturally occurring motif could be split into two pieces, an RNA target sequence and a catalytic part, the ribozyme, which could cleave the target in trans. Suitable chemical modification of the ribozyme has led to the development of oligomers which are chemically stable and nuclease resistant so that they can be used in cell culture or *in vivo*. Such chemically modified oligomers, that participate in the sequence-specific cleavage of a targeted RNA molecule, have been termed oligozymes by us.

We have developed two classes of oligozymes, one of which is based on chemically modified hammerhead ribozymes composed of 2′-O-allylribonucleotides. The latter RNA analogues possess a variety of properties which render them useful for application in molecular biology, notably chemical stability, nuclease resistance, a strong preference to bind RNA over DNA and a low level of nonspecific binding. In its most convenient form the oligozyme is a polymer of between 29 and about 36 residues whereby five residual purine ribonucleotides in the core are needed to maintain catalytic activity. The only requirement in the target RNA is an NUX site (X is not G). Although GUC and CUC sites are favoured, variations in target accessibility due to the complicated secondary and tertiary structures of the RNA may lead to better cleavage at less favoured triplets.

The other oligozyme class is based on Innovirs proprietary EGS (External Guide Sequence) technology. EGS oligozymes are small, chemically modified oligonucleotides designed to bind to specific RNA target sites to generate motifs resembling precursor tRNA, which can be cleaved by the naturally occurring cellular ribozyme RNase P.

New sequence information is appearing at a rapid rate, and the determination of the biological role of a new gene is a formidable task. The new tools described here enable one to destroy the RNA transcript from any particular gene, thereby preventing translation to protein. This is equivalent to a gene knock-out experiment. This targeted destruction of specific RNAs can of course be applied to genes thought to be involved in a particular disease; this is called drug target validation. An oligozyme that is effective in this way can in theory be regarded as a lead compound that can be further optimized for development as a therapeutic agent. Such compounds have been used successfully *in vitro*, in cell culture and *in vivo*. A particularly striking result was the use of a chemically modified hammerhead ribozyme to demonstrate the importance of amelogenin in the formation of tooth enamel in mice.

1 Introduction

The unique base sequence of an RNA should in principle enable the design of a so-called antisense sequence that binds uniquely and specifically to its target by simple Watson-Crick base pairing and so interfere with its translation into protein.

Zamecnik and Stephenson were the first to demonstrate that a short oligonucleotide complementary to a part of the Rous sarcoma virus was able to inhibit replication (Zamecnik and Stephenson, 1978). Since those early days an enormous effort has been put into identifying new oligomers capable of arresting translation of a particular RNA or leading to its destruction by the endogenous RNase H, a ubiquitous enzyme that cleaves the RNA portion of an RNA–DNA hybrid (Berkower *et al.*, 1973). The future for rationally designed drugs based on this general approach holds great promise with several oligonucleotide analogues already in various stages of clinical trials.

During the 1980s, research in the laboratories of Cech and Altman led to the discovery of certain RNAs that were able to function catalytically like enzymes (Cech *et al.*, 1981; Kruger *et al.*, 1982; Guerrier-Takada *et al.*, 1983). These molecules were termed ribozymes since they were RNA not protein based. Several classes of ribozymes have now been identified, the *Tetrahymena* group I intron (Cech *et al.*, 1981), the group II self-splicing intron (Van der Veen *et al.*, 1986; Michel *et al.*, 1989), the hammerhead ribozyme (Buzayan *et al.*, 1986; Prody *et al.*, 1986), the hairpin ribozyme (Hampel and Tritz, 1989), the hepatitis delta virus ribozyme (Been *et al.*, 1992), RNase P (Guerrier-Takada *et al.*, 1983) and a modified hammerhead structure found in the satellite 2 transcripts from newt and salamanders (Garrett *et al.*, 1996). Except for RNase P, which cleaves in trans in nature, all the other ribozymes are self-cleaving structures as originally discovered. Later it was demonstrated that by appropriate design all of the cis-cleaving ribozymes could be made to work in trans, such that an enzyme part could be made to cleave any RNA target molecule which was recognized by sequence complementarity.

The smallest of the ribozymes is the so-called hammerhead ribozyme which was discovered in nature as a structural motif in certain plant viroids, *viz.* the avocado sunblotch viroid (Hutchins *et al.*, 1986), the peach latent mosaic viroid (Hernández and Flores, 1992) and the carnation stunt associated viroid (Hernández *et al.*, 1992), and the satellite RNAs of some plant viruses, *viz.* barley yellow dwarf virus, arabis mosaic virus, chicory yellow mottle virus, lucerne transient streak virus (Forster and Symons, 1987), tobacco ring spot virus (Buzayan *et al.*, 1986; Prody *et al.*, 1986), subterranean clover mottle virus, solanum nodiflorum mottle virus and velvet tobacco mottle virus. A sequence compilation has been recently made and shows the catalytic motifs that are found in these catalytic RNAs (Bussière *et al.*, 1996). The ribozyme is involved in rolling-circle replication in these plant pathogens (Symons, 1989). The motif is illustrated in Fig. 1 using the now standard numbering system (Hertel *et al.*, 1992). The structure consists of three RNA helices and a single-stranded loop of highly conserved residues (Keese and Symons, 1987). For the sake of clarity, in describing the

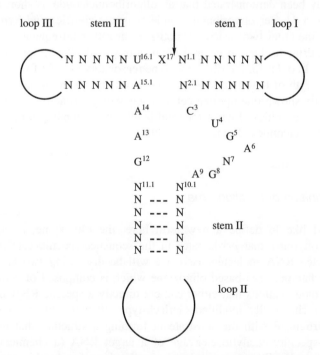

Fig. 1 Stylized version of the hammerhead ribozyme motif using the now standard numbering system (Hertel *et al.*, 1992) to show how the three different ribozymes can be derived.

different ribozymes which can be obtained from this motif the three helices have been closed with loops. It was eventually deduced that cleavage occurs after a NUX triplet, where N is any ribonucleoside and X is not guanosine (Koizumi *et al.*, 1988; Ruffner *et al.*, 1990; Perriman *et al.*, 1992). Cleavage occurs by way of magnesium ion assisted attack of the 2′-hydroxyl residue of the X ribose moiety on the adjacent phosphodiester linkage leading to the formation of the 2′,3′-cyclic phosphate of the X residue. How the hammerhead structure folds up and how an in-line displacement can occur can be seen from the recently solved X-ray crystal structures (Pley *et al.*, 1994; Scott *et al.*, 1995, 1996).

The cis-cleaving motif can be split up in a number of ways to generate systems capable of cleavage in trans. The Gerlach design (Haseloff and Gerlach, 1988), is obtained by separating the motif in Fig. 1 through stem/loops I and III and this version places the minimal requirements on the target sequence, *viz.* the presence of a NUX site, but results in a larger ribozyme than the other two possibilities. Such sites are of course very frequent in any RNA sequence. Separating the motif through stem/loops I and II leads to the Uhlenbeck ribozyme whereby part of the conserved core, the GAAA part, has to come from the target RNA (Uhlenbeck, 1987). Separating the motif through stem/loops II and III generates the smallest ribozyme of all, however the greater part of the conserved core, the CUGANGA region, has to be provided by the target RNA. Nonetheless, it has already been demonstrated that an oligoribonucleotide as short as a 13-mer can cleave a 41-mer oligoribonucleotide substrate (Jeffries and Symons, 1989). However, the latter two structural motifs occur rather infrequently in nature and do not lend themselves to general applications as do the simpler Gerlach-style ribozymes. In order for the ribozyme to be reasonably specific for a single mRNA the total length of both hybridizing arms should be about 14 bases. Moreover it was recently shown that optimal cell efficiency of 2′-O-methylated Gerlach style ribozymes was achieved when the total length of the binding arms was between 13 and 15 nucleotides (Jarvis *et al.*, 1996b).

2 Oligozymes

2.1 Definition of an oligozyme

We would like to define a new term here, the oligozyme, as a chemically modified oligomer that participates in the sequence-specific catalytic cleavage of a targeted RNA molecule. Here we will be discussing two basic versions: firstly the hammerhead-based oligozyme which is composed of certain types of chemical modifications that either can cut directly a specific RNA in a catalytic manner (a chemically modified Gerlach-type ribozyme fits in this category), or can participate with the substrate in forming a structure that results in the sequence-specific, catalytic cleavage of a target RNA (a chemically modified Uhlenbeck-style ribozyme or ultrashort ribozyme fits in this category); and secondly the external guide sequence (EGS) which is an oligozyme that binds to a targeted RNA and thereby elicits catalytic cleavage by the cellular enzyme RNase P.

2.2   Chemical modification of the hammerhead ribozyme

As we discussed above there are synthetic catalytic oligonucleotides based on the hammerhead ribozyme and for general applications of our technology we will restrict ourselves to those based on the Gerlach design derived from the positive strand of the satellite RNA of the tobacco ring spot virus (Buzayan *et al.*, 1986; Prody *et al.*, 1986). This molecule has a length of about 36 nucleotides, which is considerably longer than standard antisense molecules. The superiority of using hammerhead ribozymes compared with phosphorothioate antisense molecules to the same site with regard to efficacy and specificity has been reported (Jarvis *et al.*, 1996b); however, this may not be general. The question of whether cleavage offers an advantage compared to simple steric blocking has also been addressed in a recent review (Woolf, 1995).

Unfortunately, using the standard antisense approach in its generally accepted form one is at the mercy of RNase H with regard to the chemical modifications that are accepted. Thus current developments consist in making chimeric molecules containing a central window of phosphodiester or preferably phosphorothioate linked nucleosides surrounded by wings which confer nuclease resistance and may comprise alternative backbones but nonetheless enable RNase H to cleave the hybridized RNA (Lima and Crooke, 1997). The fact that ribozymes are catalytic in their own right leads to considerably less restrictions on the type of chemical modifications that can be introduced without loss of cleavage activity.

The greatest problem with chemically synthesized RNA is its intrinsic instability to a plethora of nucleases, in particular ribonucleases, such that in active serum or in cell culture it is rapidly destroyed. Several groups have been involved in making chemically modified ribozymes that can be used for experiments in cell culture or *in vivo*.

Thus it proved necessary as a minimum requirement to replace all the ribopyrimidines, as these seemed to be a major weak site of attack for RNase A-like activity (Heidenreich and Eckstein, 1992). Thus Eckstein and co-workers have synthesized chemically modified hammerhead ribozymes in which many of the ribopyrimidines and/or ribopurines were replaced by their corresponding $2'$-deoxy-$2'$-fluoro or $2'$-deoxy-$2'$-amino analogues (Olsen *et al.*, 1991; Pieken *et al.*, 1991; Williams *et al.*, 1992). Use of these analogues plus a couple of phosphorothioate linkages at each end afforded modified ribozymes with excellent activity and good nuclease resistance (Heidenreich *et al.*, 1994). Other groups have concentrated their efforts on synthesizing chemically modified hammerhead ribozymes in which most of the ribonucleotides are replaced by $2'$-O-methylribonucleotides (Goodchild, 1992; Yang *et al.*, 1992; Usman *et al.*, 1994), and this has led to the identification of highly active structures containing a minimum of five key ribonucleotides in the conserved core region necessary for catalytic activity (Usman et al., 1994). Our own efforts are based on hammerhead structures comprising mostly 2'-O-allylribonucleotides (Paolella et al., 1992), which we now call oligozymes.

2.3 Derivation of the external guide sequence

RNase P is a ubiquitous enzyme that is involved in the maturation of all pre-tRNAs (Guerrier-Takada et al., 1983). The enzyme is found in the nucleus in eukaryotes, however there is also evidence for active enzyme in the mitochondria (Doersen et al., 1985). It was eventually shown that the structure recognized and processed by bacterial RNase P could be substantially reduced in size to a molecule resembling the T-stem and loop stacked on the acceptor stem of a tRNA (McClain et al., 1987). Further reduction in the size of the substrate was achieved when it was discovered that the enzyme did not require the T-loop in the previous structure and would in fact cleave a target RNA molecule when this was base-paired with an external guide sequence (EGS) to generate a structure mimicking part of a pre-mRNA (Forster and Altman, 1990). Finally it was demonstrated that chemically synthesized external guide sequences enabled bacterial RNase P to cleave a matching substrate RNA in vitro (Li et al., 1992).

Unfortunately the mammalian enzyme turned out to be more fussy than the bacterial enzyme and required a much larger external guide sequence to force it to cut a target RNA sequence (Yuan et al., 1992; Yuan and Altman, 1994), in fact representing about three quarters of a tRNA. The EGS had a length of about 63 nucleotides which is a challenge for chemical synthesis. Further systematic deletion analysis of the very long EGS showed that this could be reduced in size to a molecule comprising the acceptor stem plus T-stem and loop and a D-stem (S. T. George, M. Werner and A. R. Goldberg; personal communication). The features of this much shortened EGS are illustrated in Fig. 2. In this case the EGS is based on the tyrosyl-tRNA precursor. The following features are important: The cleavage site is usually G, but can be N, and a bulge of flexible sequence with a length of one to three nucleotides is required in the target RNA opposite the T-stem. The acceptor stem and D-stem can have flexible sequences with optimal lengths of seven and four to nine base pairs respectively. The T-stem is a known tRNA sequence and has an optimal length of five base pairs, likewise the T-loop is a known tRNA sequence and the length must be seven nucleotides. This generates an EGS of length about 28–32 nucleotides, which can easily be synthesized chemically. In order for exogenously administered EGSs to be useful they must be chemically modified to prevent degradation by RNases. The best combination of chemical modifications that led to reasonable cleavage activity and high stability as measured in 50% human serum was to use 2'-O-methylribonucleoside phosphorothioates in the two terminal positions at the 3'- and 5'-ends of the EGS, with the remainder of the molecule composed of 2'-O-methyl RNA except for the critical T-loop which comprised 2'-O-methylribonucleotides at pyrimidine positions and ribonucleotides at the purine positions. These modifications led to a

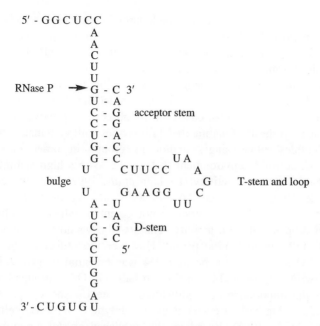

Fig. 2 Structural features of the shortened EGS based on the tyrosyl-tRNA precursor hybridized to its target RNA and showing the RNase P cleavage position.

half life of about 18 h, compared to less than 10 min for the unmodified EGS. These chemically modified EGSs constitute the second class of oligozymes as defined above (M. Ma and S. T. George, personal communication).

3 Oligozymes based on the hammerhead ribozyme

3.1 Controlling RNA instability by chemical modification

We developed oligo(2'-O-alkylribonucleotide)s as special probes for investigating RNA processing (Iribarren *et al.*, 1990) and for a variety of antisense experiments (Lamond and Sproat, 1993). Highly desirable properties were that the probe should bind tightly and stably to RNA, it should be chemically stable, nuclease-resistant, show minimal non-specific binding interactions particularly with proteins and hybrids with targeted RNA should not be substrates for RNase H. These properties were largely fulfilled by the 2'-O-methyl analogues; however with regard to nuclease resistance and reduced non-specific interactions with nuclear proteins the 2'-O-allyl (the correct chemical name for allyl is prop-2-enyl) analogues which we developed proved greatly superior (Iribarren *et al.*, 1990). More recently, other groups have identified alternative compounds of this general 2'-O-alkylRNA class with interesting characteristics (Lima and Crooke, 1997). Thus, oligo(2'-O-allylribonucleotide)s proved very useful for the isolation of specific small nuclear ribonucleoprotein particles (snRNPs) involved in mammalian pre-mRNA splicing (Lamm *et al.*, 1991), and moreover when tagged with fluorophores could be used for locating specific snRNPs in the nuclei of living cells (Carmo-Fonseca *et al.*, 1991).

We became interested in the chemical modification of hammerhead ribozymes to make tools that could be used in the molecular biology laboratory to investigate gene function. Thus, based on the observations above, we embarked on the chemical synthesis of hammerhead ribozymes with systematic replacement of ribonucleotides either singly and/or in groups in order to identify a minimum ribonucleotide environment that would possess high stability and yet retain substantial catalytic activity (Paolella *et al.*, 1992). We showed that the ribonucleotides at positions G5, A6, G8, G12 and A15.1 were essential for catalytic activity, and that U4 although not essential, when 2'-O-allylated led to a marked drop in catalytic activity. The six-ribo version was about 20% as active as its full ribonucleotide parent. However the utilization of molecules containing 2'-O-allyluridine at position U4 was essential to give full stability against RNase A activity and a long serum half-life. Other groups have shown in particular the importance of modifications at the U4 and U7 positions with regard to maintaining high or even enhanced catalytic activity (Beigelman *et al.*, 1995; Burgin *et al.*, 1996). Based on the published crystal structures and on the kinetics data of several groups, a picture emerges suggesting that the best tolerated analogues at position U4 should be those uridine analogues which can adopt or prefer the C2'-endo sugar pucker, which would explain the results with 2'-deoxyuridine (Tanaka *et al.*, 1994), 2'-C-allyluridine (Beigelman *et al.*, 1995) and 2'-amino-2'-deoxyuridine (Beigelman *et al.*, 1995). Analogues such as 2'-O-methyl-and 2'-O-allyl-uridine which are predominantly in the 3'-endo form function less well. Thus we were able to produce a heavily substituted molecule which behaved similarly to RNA in terms of overall structure, but had greatly enhanced chemical stability and nuclease resistance as well as reduced non-specific interactions with nuclear proteins. More recently it has been shown that a substantial part of stem II and the entire loop II can be replaced by simple linker molecules such as hexa(ethylene glycol), resulting in shorter, higher yield syntheses (Benseler *et al.*, 1993; Thomson *et al.*, 1993; Fu *et al.*, 1994).

3.2   The chemical synthesis of oligozymes based on the hammerhead ribozyme

2'-O-Allyl modified Gerlach type hammerhead ribozymes containing five residual purine ribonucleotides (G5, A6, G8, G12 and A15.1) bearing a suitable 3'-terminus such as an inverted thymidine residue (Ortigao *et al.*, 1992) or two phosphorothioate linkages at the 3'-terminus to prevent eventual degradation by 3'-exonucleases, can be synthesized by solid phase β-cyanoethyl phosphoramidite chemistry (Sinha *et al.*, 1984) on any commercially available DNA/RNA synthesizer. We use the 2'-O-*tert*-butyldimethylsilyl (TBDMS) protection strategy for the ribonucleotides (Usman *et al.*, 1987) and all the required 3'-O-phosphoramidites are commercially available. In addition we favour the use of aminomethylpolystyrene as the support material due to its advantageous properties (McCollum and Andrus, 1991). The desired oligozyme is synthesized using a standard RNA cycle. Upon completion of the assembly all base labile protecting groups are removed by an 8 h treatment at 55°C with concentrated aqueous ammonia/ethanol (3:1 v/v) in a sealed vial. The ethanol suppresses premature removal of the 2'-O-TBDMS groups which would otherwise lead to appreciable strand cleavage at the resulting ribonucleotide positions under the basic conditions of the deprotection (Usman *et al.*, 1987). After lyophilization the TBDMS protected oligozyme is treated with a mixture of triethylamine trihydrofluoride/triethylamine/N-methylpyrrolidinone for 2 h at 60°C to afford fast and efficient removal of the silyl protecting groups under neutral conditions (Wincott *et al.*, 1995). The fully deprotected oligozyme is then precipitated with butanol according to the published procedure (Cathala and Brunel, 1990). Purification can either be performed by denaturing polyacrylamide electrophoresis or by a combination of ion-exchange HPLC (Sproat *et al.*, 1995) and reversed phase HPLC. To be used for biological work all oligozymes are converted to their sodium salts by precipitation with sodium perchlorate in acetone. Traces of residual salts are then best removed using small disposable gel filtration columns that are commercially available. As a final step it is recommended to check the authenticity of the isolated oligozymes by MALDI-ToF mass spectrometry (Pieles *et al.*, 1993) and by nucleoside base composition analysis. In addition, we usually perform a functional cleavage test with the oligozyme on the corresponding chemically synthesized short oligoribonucleotide substrate.

4   Choosing a target site on an RNA

4.1   The importance of locating a good target site

Locating a suitable target site on a particular mRNA is not an easy proposition, regardless of whether one is intending to use antisense or oligozyme technology. Ideally one would like to find regions of the mRNA that are accessible, for instance single stranded regions or large loops, such that the appropriate oligozyme can anneal rapidly and tightly. This is particularly important since the rate-limiting step for cleavage of a long RNA within a cell will almost certainly be the formation of the active oligozyme–RNA substrate complex and not the chemical step. Moreover, judicious choice of a good target site will therefore be expected to dramatically improve the specificity of any particular oligozyme.

4.2   Locating a good target site

RNA is of course a very dynamic structure and during the various processing events that occur post-transcriptionally, such as polyadenylation and splicing in the cell nucleus, a large number of sequence-specific and sequence non-specific protein factors will be bound to it, thus rendering regions of the RNA inaccessible in a time-dependent manner. The mature mRNA is then presumably actively transported through the nuclear pore complex into the cytoplasm to be ultimately translated on the ribosome where a plethora of other protein factors bind to it. Many of these protein factors have an associated helicase activity, thus unwinding double stranded regions to enable translation to take place. Indeed many of those antisense compounds that do not elicit RNase H cleavage of the RNA target get brushed aside by the translational machinery when hybridized to the coding region of the RNA.

Large RNAs can be examined using the Zuker MFOLD algorithm (Zuker, 1989) to give some idea where there may be little secondary structure; however one should not place a great deal of reliability on this method which views the RNA as a static molecule. At present we synthesize a selection of oligozymes against GUC sites as these generally have the best cleavage kinetics *in vitro* (Shimayama *et al.*, 1995; Zoumadakis and Tabler, 1995) although the cleavage rate *in vivo* will depend more on the surrounding sequences (Sullivan, 1994) and the site accessibility. Those oligozymes which have serious secondary structures are omitted. It is then wise to synthesize a reasonable selection of oligozymes predicted to have little or no secondary structure and to screen them against the desired target RNA in cell culture or in a cell lysate. In this way one can rapidly identify oligozymes that should be good enough for a target validation experiment. In some cases accessible regions on a particular RNA may already be known from the literature but for new sequence data this will not be the case.

There exist two alternative possibilities since one cannot yet predict RNA accessibility by computer. Screening a particular target RNA with a range of anti-sense oligonucleotides covering the entire sequence in combination with RNase H will enable the identification of highly accessible sites, which will be particularly important if one is trying to identify a lead compound for further development as a therapeutic. This approach has been used to locate accessible ribozyme sites on murine c-*myb* mRNA (Jarvis *et al.*, 1996a). Moreover, a chemical library approach was successful in locating potent antisense oligonucleotides to the human multiple drug resistance mRNA (Ho *et al.*, 1996). It has already been demonstrated that this method is much more reliable than MFOLD in locating good sites for ribozyme cleavage (Birikh *et al.*, 1997). A rather elegant approach using a fully randomized decamer to locate accessible sites in the 5′-terminal region of the hepatitis C genomic RNA proved very successful (Lima *et al.*, 1997) and is of course the logical extension to using a limited set of antisense sequences. Nonetheless, as a word of precaution, if proteins are bound to the RNA in the vicinity of an otherwise accessible site then RNase H may not be able to function due to steric hindrance, even though an oligozyme could cleave at that site.

Ideally one would like a generic method that works in cell culture so that the RNA is at least dynamic. To this end a transcribed ribozyme approach has been developed whereby the ribozyme has randomized hybridizing arms such that all possible sequences are represented (Lieber and Strauss, 1995). It is clear that many of these methods have either experimental or time and expense limitations, however as more experimental data accumulates and we understand more about RNA structure the rate of target site predictability by computer should increase dramatically. At present, computer-derived structures of large RNA molecules should be viewed with caution.

Considering all the above problems, although it might prove difficult or impossible to cleave a long RNA transcript *in vitro* since the molecule is static and will in any case consist of a population of structures of similar energies in which the desired cleavage site may well be inaccessible, the situation in the cell or living animal may well not be so bad. Several proteins, namely the p7 nucleo-capsid protein of human immunodeficiency virus type-1 and the heterogenous nuclear ribonucleoprotein A1, have been demonstrated to facilitate the catalysis of hammerhead ribozymes (Tsuchihashi *et al.*, 1993; Bertrand and Rossi, 1994; Herschlag *et al.*, 1994; Müller *et al.*, 1994). The effects of these non-specific binding proteins is to open up RNA secondary structure thereby increasing the annealing rate of the ribozyme and increasing the product release. On the other hand there may be differential effects when differently chemically modified ribozymes are used for experiments in cells as well as differential partitioning between the nucleus and cytoplasm. If the ribozyme–substrate complex is recognized by the non-specific binding proteins then the dissociation rate may exceed the rate of the chemical cleavage when the hybridizing arms are short, which may well explain recently published differences in efficacy of ribozymes in human cells (Hormes *et al.*, 1997). We strongly suspect that hybrids of 2'-O-allyl modified ribozymes and their target RNAs are only poorly recognized by these proteins, which implies that even highly modified constructions with correspondingly reduced cleavage characteristics should be relatively efficient in cleaving long RNA targets especially in the nucleus. This of course needs to be confirmed by appropriate experiments; however, our own animal experiments suggest that this is probably the case.

5 Elucidating gene function *in vivo*

5.1 The role of amelogenins

Amelogenins are thought to play an important role in mammalian dental enamel biomineralization by acting as a matrix to control the crystallization of hydroxy-apatite crystals (Fincham *et al.*, 1994). A diversity of amelogenins are produced by alternative splicing of the mouse amelogenin pre-mRNA (Lau *et al.*, 1992), as the major product of ameloblasts, which are single-layered columnar epithelial cells lining the crown of the developing tooth (Termine *et al.*, 1980). An examination of the mouse amelogenin mRNA sequence showed five GUC sites. We initially synthesized a 2'-O-allyl modified hammerhead ribozyme containing five residual ribonucleotides (G5, A6, G8, G12 and A15.1) targeted to the first GUC site in the coding region of the mRNA and carrying either a 3'-inverted thymidine (Ortigao *et al.*, 1992) or two 3'-terminal phosphorothioate linkages to prevent degradation by 3'-exonucleases *in vivo* (Lyngstadaas *et al.*, 1995). As control an inactive ribozyme carrying a G12 to A12 mutation and a standard oligo(2'-O-allylribonucleotide) antisense molecule were synthesized.

The chemically modified active oligozymes (25 or 50 μg), the inactive control, the antisense molecule or saline were injected into the jaws of newborn mice on the lingual side between the developing first and second molars (Lyngstadaas *et al.*, 1995). At various time points up to four days starting at time zero a four-hour pulse of [^{35}S]-methionine was given by intraperitoneal injection. The mice were then sacrificed and the first molar tooth buds were removed and the proteins were analysed on a 12% SDS-polyacrylamide gel and quantitated by phosphor-imaging. At a dose of 50 μg of active oligozyme per animal a complete arrest of amelogenin synthesis was observed (the bands at 22, 23, 25 and 27 kDa all disappeared relative to the control) for 24 hours. Even after three days the level had only recovered to about 50%. Moreover, since all protein bands disappeared and the target site is in an exon which is not present in all transcripts, then it would appear that the oligozyme is functioning in the cell nucleus and cleaving the pre-mRNA. This is not unexpected as similar results on other targets have been observed using antisense technology (Condon and Bennett, 1996). Bands due to housekeeping genes were not affected. It took four days for the amelogenin synthesis to return to normal during which time the oligozyme was presumably degraded by RNase activity. This is the eventual fate of all chemically modi-fied ribozymes, even those containing a small number of largely non-contiguous purine ribonucleotides (Beigelman *et al.*, 1995). The mutated oligozyme and the antisense control showed a transient effect on translation which lasted a few hours only and was presumably a non-specific effect caused by the local high concentration immediately after the injection, since it is known that oligo(2'-O-allylribonucleotide)s targeted against the coding region of a mRNA are not able to prevent translation (Johansson *et al.*, 1994).

In order to examine the effect of upsetting amelogenin synthesis for a substan-tial time (the gene is switched on for about eight days after birth), teeth from five-week-old mice that had been injected once with oligozyme on the first day after birth were examined by scanning transmission electron microscopy. Whereas control mice showed perfectly normal dental enamel with correctly ordered arrays of hydroxyapatite crystals, the enamel of the oligozyme treated mice revealed severe hypomineralization with no visible crystals within cross-cut prism domains. The second molars were more severely affected than the first molars reflecting the importance of amelogenin in the very early stage of enamel development. The eruption time of the teeth was not affected nor was the general morphology. Ribozymes targeted to the other four GUC sites were not as effective, and the inac-tive controls often showed substantial inhibition. This is not particularly surprising as several of these sequences contained G quartets at the ends.

This experiment shows that oligozymes do work in animal experiments even without cell permeators. Moreover the results confirm the key role played by amelogenins in tooth enamel development, *viz.* these secreted proteins function as a matrix for organizing the hydroxyapatite crystallization. Upon completion of enamel maturation the amelogenin matrix is digested by a specific protein to leave the crystalline hydroxyapatite and structural proteins such as enamelins (Moradian-Oldak *et al.*, 1996).

5.2 Timed knockout of amelogenin

The newborn mouse exhibits three differently staged molar teeth and one incisor in each jaw. The incisor, however, is continuously growing, exhibiting all stages of tooth formation, each confined to a specific area along the tooth. In the case of the first molar the amelogenin gene is switched on for eight days after birth. Injection of active oligozymes on day one or two resulted in seriously damaged enamel in the mature erupted first molars. Injection on day six had much less effect and on day eight there was no effect on amelogenin synthesis due to the completion of the secretory stage.

Regarding second molars, the amelogenin gene is switched on from day two to day twelve after birth. Timing the injection to the late secretory stage caused only the cervical parts of the enamel to be hypomineralized. Injections on days six and eight had no effect on the third molars which secrete amelogenins from days nine to nineteen after birth and which erupt about four weeks after birth. In the incisors the expression of amelogenin was stopped for 12 hours after injection but recovered fully within 72 hours and the effect on the dental enamel as judged by electron microscopy was minimal.

The above experiments nicely demonstrate how oligozymes can be used as a new tool in molecular biology to identify or confirm the function of genes using timed knock-outs.

6 Conclusion

The technology described here can be used to identify the function of new genes. The real potential, however, lies in the validation of drug targets for the pharmaceutical industry. The idea would be to knock out a particular gene whose presence or overexpression is responsible for a particular disease state. The use of ribozymes for the inhibition of gene expression and the perspectives for therapeutic applications have been reviewed recently (Christoffersen and Marr, 1995; Kijima *et al.*, 1995; Marschall *et al.*, 1994; Usman and McSwiggen, 1995) and cover not only chemically synthesized ribozymes but also transcribed ribozymes.

In the long term it is hoped to develop oligozymes based on external guide sequences and hammerhead ribozymes as a new class of rationally designed therapeutics. Results in an animal model of arthritis have shown activity of 2′-O-methyl hammerhead ribozymes against the stromelysin mRNA (Flory *et al.*,

1996) and we soon expect to obtain further *in vivo* results with targets involved in cancer. Although it will be a challenge to chemically modify oligozymes to obtain organ- or cell-specific targeting, 2'-O-allyl modified ribozymes delivered by simple intravenous injection in saline in rats do show a reasonably good biodistribution (Desjardins *et al.*, 1996).

References

Been, M. D., Perrotta, A. T. and Rosenstein, S. P. (1992). *Biochemistry* **31**, 11843–11852.
Beigelman, L., McSwiggen, J., Draper, K., Gonzalez, C., Jensen, K., Karpeisky, A., Modak, A., Matulic-Adamic, J., DiRenzo, A., Haeberli, P., Sweedler, D., Tracz, D., Grimm, S., Wincott, F., Thrackray, V. and Usman, N. (1995). *J. Biol. Chem.* **270**, 25702–25708.
Benseler, F., Fu, D., Ludwig, J. and McLaughlin, L. W. (1993). *J. Am. Chem. Soc.* **115**, 8483–8484.
Berkower, I., Leis, J. and Hurwitz, J. (1973). *J. Biol. Chem.* **248**, 5914–5921.
Bertrand, E. L. and Rossi, J. J. (1994). *EMBO J.* **13**, 2904–2912.
Birikh, K. R., Berlin, Y. A., Soreq, H. and Eckstein, F. (1997). *RNA* **3**, 429–437.
Burgin Jr., A. B., Gonzalez, C., Matulic-Adamic, J., Karpeisky, A. M., Usman, N., McSwiggen, J. A. and Beigelman, L. (1996). *Biochemistry* **35**, 14090–14097.
Bussière, F., Lafontaine, D. and Perreault, J.-O. (1996). *Nucleic Acids Res.* **24**, 1793–1798.
Buzayan, J. M., Gerlach, W. L. and Bruening, G. (1986). *Nature* **323**, 349–353.
Cathala, G. and Brunel, C. (1990). *Nucleic Acids Res.* **18**, 201.
Carmo-Fonseca, M., Pepperkok, R., Sproat, B. S., Ansorge, W., Swanson, M. S. and Lamond, A. I. (1991). *EMBO J.* **10**, 1863–1873.
Cech, T. R., Zaug, A. J. and Grabowski, P. J. (1981). *Cell* **27**, 487–496.
Christoffersen, R. E. and Marr, J. J. (1995). *J. Med. Chem.* **38**, 2023–2037.
Condon, T. P. and Bennett, C. F. (1996). *J. Biol. Chem.* **271**, 30398–30403.
Desjardins, J. P., Sproat, B. S., Beijer, B., Blaschke, M., Dunkel, M., Gerdes, W., Ludwig, J., Reither, V., Rupp, T. and Iversen, P. L. (1996). *J. Pharmacol. Exp. Ther.* **278**, 1419–1427.
Doersen, C.-J., Guerrier-Takada, C., Altman, S. and Attardi, G. (1985). *J. Biol. Chem.* **260**, 5942–5949.
Fincham, A. G., Moradian-Oldak, J., Simmes, J. P., Sarte, P., Lau, E. C., Diekwish, T. and Slavkin, H. C. (1994). *J. Struct. Biol.* **112**, 103–109.
Flory, C. M., Pavco, P. A., Jarvis, T. C., Lesch, M. E., Wincott, F. E., Beigelman, L., Hunt III, S. W. and Schrier, D. J. (1996). *Proc. Natl. Acad. Sci. USA* **93**, 754–758.
Forster, A. C. and Altman, S. (1990). *Science* **249**, 783–786.
Forster, A. C. and Symons, R. H. (1987). *Cell* **49**, 211–220.
Fu, D., Benseler, F. and McLaughlin, L. W. (1994). *J. Am. Chem. Soc.* **116**, 4591–4598.
Garrett, T. A., Pabón-Peña, L. M., Gokaldas, N. and Epstein, L. M. (1996). *RNA* **2**, 699–706.
Goodchild, J. (1992). *Nucleic Acids Res.* **20**, 4607–4612.
Guerrier-Takada, C., Gardiner, K., Marsh, T., Pace, N. and Altman, S. (1983). *Cell* **35**, 849–857.
Hampel, A. and Tritz, R. (1989). *Biochemistry* **28**, 4929–4933.
Haseloff, J. and Gerlach, W. (1988). *Nature* **334**, 585–591.
Heidenreich, O. and Eckstein, F. (1992). *J. Biol. Chem.* **267**, 1904–1909.
Heidenreich, O., Benseler, F., Fahrenholz, A. and Eckstein, F. (1994). *J. Biol. Chem.* **269**, 2131–2138.
Hernández, C., Daròs, J. A., Elena, S. F., Moya, A. and Flores, R. (1992). *Nucleic Acids Res.* **20**, 6323–6329.

Hernández, C. and Flores, R. (1992). *Proc. Natl. Acad. Sci. USA* **89**, 3711-3715.

Herschlag, D., Khosla, M., Tsuchihashi, Z. and Karpel, R. L. (1994). *EMBO J.* **13**, 2913-2924.

Hertel, K. J., Pardi, A., Uhlenbeck, O. C., Koizumi, M., Ohtsuka, E., Uesugi, S., Cedergren, R., Eckstein, F., Gerlach, W. L., Hodgson, R. and Symons, R. H. (1992). *Nucleic Acids Res.* **20**, 3252.

Ho, S. P., Britton, D. H. O., Stone, B. A., Behrens, D. L., Leffet, L. M., Hobbs, F. W., Miller, J. A. and Trainor, G. L. (1996). *Nucleic Acids Res.* **24**, 1901-1907.

Hormes, R., Homann, M., Oelze, I., Marschall, P., Tabler, M., Eckstein, F. and Sczakiel, G. (1997). *Nucleic Acids Res.* **25**, 769-775.

Hutchins, C. J., Rathjen, P. D., Forster, A. C. and Symons, R. H. (1986). *Nucleic Acids Res.* **14**, 3627-3640.

Iribarren, A. M., Sproat, B. S., Neuner, P., Sulston, I., Ryder, U. and Lamond, A. I. (1990). *Proc. Natl. Acad. Sci. USA* **87**, 7747-7751.

Jarvis, T. C., Alby, L. J., Beaudry, A. A., Wincott, F. E., Beigelman, L., McSwiggen, J. A., Usman, N. and Stinchcomb, D. T. (1996a). *RNA* **2**, 419-428.

Jarvis, T. C., Wincott, F. E., Alby, L. J., McSwiggen, J. A., Beigelman, L., Gustofson, J., DiRenzo, A., Levy, K., Arthur, M., Matulic-Adamic, J., Karpeisky, A., Gonzalez, C., Woolf, T. M., Usman, N. and Stinchcomb, D. T. (1996b). *J. Biol. Chem.* **271**, 29107-29112.

Jeffries, A. C. and Symons, R. H. (1989). *Nucleic Acids Res.* **17**, 1371-1377.

Johansson, H. E., Belsham, G. J., Sproat, B. S. and Hentze, M. W. (1994). *Nucleic Acids Res.* **22**, 4591-4598.

Keese, P. and Symons, R. H. (1987). In *Viroids and viroid-like pathogens* (J. S. Semancik, ed.), pp 1-47. CRC Press, Boca Raton, Florida.

Kijima, H., Ishida, H., Ohkawa, T., Kashani-Sabet, M. and Scanlon, K. J. (1995). *Pharmacol. Ther.* **68**, 247-267.

Koizumi, M., Iwai, S. and Ohtsuka, E. (1988). *FEBS Lett.* **228**, 228-230.

Kruger, K., Grabowski, P. J., Zaug, A. G., Sands, J., Gottschling, D. F. and Cech, T. R. (1982). *Cell* **31**, 147-157.

Lamm, G. M., Blencowe, B. J., Sproat, B. S., Iribarren, A. M., Ryder, U. and Lamond, A. I. (1991). *Nucleic Acids Res.* **19**, 3193-3197.

Lamond, A. I. and Sproat, B. S. (1993). *FEBS Lett.* **325**, 123-127.

Lau, E. C., Simmer, J. P., Bringas, P. Jr, Hsu, D. D.-J., Hu, C.-C., Zeichner-David, M., Thiemann, F., Snead, M. L., Slavkin, H. C. and Fincham, A. G. (1992). *Biochem. Biophys. Res. Commun.* **188**, 1253-1260.

Li, Y. C., Guerrier-Takada, C. and Altman, S. (1992). *Proc. Natl. Acad. Sci. USA* **89**, 3185-3189.

Lieber, A. and Strauss, M. (1995). *Mol. Cell. Biol.* **15**, 540-551.

Lima, W. F., Brown-Driver, V., Fox, M., Hanecak, R. and Bruice, T. C. (1997). *J. Biol. Chem.* **272**, 626-638.

Lima, W. F. and Crooke, S. T. (1997). *Biochemistry* **36**, 390-398.

Lyngstadaas, S. P., Risnes, S., Sproat, B. S., Thrane, P. S. and Prydz, H. P. (1995). *EMBO J.* **14**, 5224-5229.

Marschall, P., Thomson, J. B. and Eckstein, F. (1994). *Cell. Mol. Neurobiol.* **14**, 523-538.

McClain, W. H., Guerrier-Takada, C. and Altman, S. (1987). *Science* **238**, 527-530.

McCollum, C. and Andrus, A. (1991). *Tetrahedron Lett.* **32**, 4069-4072.

Michel, F., Umesono, K. and Ozeki, H. (1989). *Gene* **82**, 5-30.

Moradian-Oldak, J., Leung, W., Simmer, J. P., Zeichner-David, M. and Fincham, A. G. (1996). *Biochem. J.* **318**, 1015-1021.

Müller, G., Strack, B., Dannull, J., Sproat, B. S., Surovoy, A., Jung, G. and Moelling, K. (1994). *J. Mol. Biol.* **242**, 422-429.

Olsen, D. B., Benseler, F., Aurup, H., Pieken, W. A. and Eckstein, F. (1991). *Biochemistry* **30**, 9735-9741.

Ortigao, J. F. R., Rösch, H., Selter, H., Fröhlich, A., Lorenz, A., Montenarh, M. and Seliger, H. (1992). *Antisense Res. Dev.* **2**, 129-146.

Paolella, G., Sproat, B. S. and Lamond, A. I. (1992). *EMBO J.* **11**, 1913-1919.

Perriman, R., Delves, A. and Gerlach, W. L. (1992). *Gene* **113**, 157-163.

Pieles, U., Zürcher, W., Schär, M. and Moser, H. E. (1993). *Nucleic Acids Res.* **21**, 3191-3196.

Pieken, W. A., Olsen, D. B., Benseler, F., Aurup, H. and Eckstein, F. (1991). *Science* **253**, 314-317.

Pley, H. W., Flaherty, K. W. and McKay, D. B. (1994). *Nature* **372**, 68-74.

Prody, G. A., Bakos, J. T., Buzayan, J. M., Schneider, I. R. and Bruening, G. (1986). *Science* **231**, 1577-1580.

Ruffner, D. E., Stormo, G. D. and Uhlenbeck, O. C. (1990). *Biochemistry* **29**, 10695-10702.

Scott, W., Finch, J. and Klug, A. (1995). *Cell* **81**, 991-1002.

Scott, W. G., Murray, J. B., Arnold, J. R. P., Stoddard, B. L. and Klug, A. (1996). *Science* **274**, 2065-2069.

Shimayama, T., Nishikawa, S. and Taira, K. (1995). *Biochemistry* **34**, 3649-3654.

Sinha, N. D., Biernat, J., McManus, J. and Köster, H. (1984) *Nucleic Acids Res.* **12**, 4539-4557.

Sproat, B., Colonna, F., Mullah, B., Tsou, D., Andrus, A., Hampel, A. and Vinayak, R. (1995). *Nucleosides & Nucleotides* **14**, 255-273.

Sullivan, S. M. (1994). *J. Invest. Dermatol.* **103**, 85S-89S.

Symons, R. H. (1989). *Trends Biochem. Sci.* **14**, 445-450.

Tanaka, H., Endo, T., Hosaka, H., Takai, K., Yokoyama, S. and Takaku, H. (1994). *Bioorg. Med. Chem. Lett.* **4**, 2857-2862.

Termine, J. D., Belcourt, A. B., Christner, P. J., Conn, K. M. and Nylen, M. U. (1980). *J. Biol. Chem.* **225**, 9760-9768.

Thomson, J. B., Tuschl, T. and Eckstein, F. (1993). *Nucleic Acids Res.* **21**, 5600-5603.

Tsuchihashi, Z., Khosla, M. and Herschlag, D. (1993). *Science* **262**, 99-102.

Uhlenbeck, O. C. (1987). *Nature* **328**, 596-600.

Usman, N., Ogilvie, K. K., Jiang, M.-Y. and Cedergren, R. J. (1987). *J. Am. Chem. Soc.* **109**, 7845-7854.

Usman, N., Beigelman, L., Draper, K., Gonzalez, C., Jensen, K., Karpeisky, A., Modak, A., Matulic-Adamic, J., DiRenzo, A., Haeberli, P., Tracz, D., Grimm, S., Wincott, F. and McSwiggen, J. (1994). *Nucl. Acids Symp. Ser.* **31**, 163-164.

Usman, N. and McSwiggen, J. A. (1995). In *Annual Reports in Medicinal Chemistry* (J. Venuti, ed.), pp 285-294. Academic Press, New York.

Van der Veen, R. T., Arnberg, A. C., Van der Horst, G., Bonen, L., Tabak, H. F. and Grivell, L. A. (1986). *Cell* **44**, 225-234.

Williams, D. M., Pieken, W. A. and Eckstein, F. (1992). *Proc. Natl. Acad. Sci. USA* **89**, 918-921.

Wincott, F., DiRenzo, A., Shaffer, C., Grimm, S., Tracz, D., Workman, C., Sweedler, D., Gonzalez, C., Scaringe, S. and Usman, N. (1995). *Nucleic Acids Res.* **23**, 2677-2684.

Woolf, T. M. (1995). *Antisense Res. Dev.* **5**, 227-232.

Yang, J.-H., Usman, N., Chartrand, P. and Cedergren, R. (1992). *Biochemistry* **31**, 5005-5009.

Yuan, Y., Hwang, E.-S. and Altman, S. (1992). *Proc. Natl. Acad. Sci. USA* **89**, 8006-8010.

Yuan, Y. and Altman, S. (1994). *Science* **263**, 1269-1273.

Zamecnik, P. C. and Stephenson, M. L. (1978). *Proc. Natl. Acad. Sci. USA* **75**, 280-284.

Zoumadakis, M. and Tabler, M. (1995). *Nucleic Acids Res.* **23**, 1192-1196.

Zuker, M. (1989). *Science* **244**, 48-52.

11

Progress towards Therapeutic Application of SELEX-derived Aptamers

BARRY POLISKY

NeXstar Pharmaceuticals, Inc., 2860 Wilderness Place, Boulder, Colorado 80301, USA

Abstract

SELEX is a combinatorial ligand discovery technology that utilizes large libraries of random sequence oligonucleotides to isolate ligands with high affinity and specificity to molecular targets. These ligands are known as aptamers. SELEX has been used to isolate aptamers to a broad range of targets ranging from small organic molecules such as theophylline to proteins involved in pathological processes. Libraries are commonly composed of either DNA or modified RNA. An RNA modification conferring extensive *in vivo* stability relative to unmodified RNA is substitution of the 2′-OH with 2′-F on pyrimidine nucleotides. Such libraries are chemically stable upon prolonged incubation with a variety of biological sources including human plasma. Affinities of selected aptamers for growth factors and certain cell adhesion molecules are typically in the low nM to pM Kd range. Nuclease-stable aptamer antagonists to human growth factors responsible for neo-angiogenesis such as VEGF, bFGF and PDGF, and to cell adhesion molecules such as L- and P-selectin, have been isolated and tested in a variety of *in vitro* and *in vivo* model systems. Aptamers to the growth factors block their interaction with cognate receptors on cell surfaces in cell culture. Aptamers to selectins have been shown to recognize the carbohydrate domains and bind to selectins on cell surfaces. Methodology to identify specific contact positions between aptamers and targets has been developed utilizing both photochemistry and solution chemistry. Examples of both types are presented. Interactions of specific aptamers with cognate targets are described as well as progress in defining higher order structural motifs of aptamer-target complexes.

THE MANY FACES OF RNA
ISBN 0-12-233210-5

1 Introduction

SELEX (*S*ystematic *E*volution of *L*igands by *EX*ponential enrichment) is a ligand discovery procedure that uses large libraries of random sequence single-stranded oligonucleotides composed of either RNA, DNA or modified nucleotides (Tuerk and Gold, 1990; Ellington and Szostak, 1990). RNA libraries are produced by transcription of double-stranded DNA templates containing stretches of sequence that were randomized during solid phase synthesis. These oligonucleotide libraries are exposed to the target of interest, and subsets of sequences that have affinity are partitioned from those sequences with little or no affinity. SELEX is an iterative process in which the bound oligonucleotides are extracted from the target, converted to double-stranded DNA and amplified by PCR (Fig. 1). Following transcription of the amplified population, another round of exposure to the target is carried out. Each round of SELEX takes about a day to do. Typically, $10-15$ rounds of SELEX are carried out under biochemical conditions designed to create stringent competition among RNA populations for the subset of sequences that have the highest affinity for the target. These rounds result in a reduction of sequence complexity of the initial library from 10^{14} to roughly $10^1 - 10^3$ depending on the target and the conditions used. These sequences are then cloned into bacterial plasmids and their primary sequences determined and aligned. Usually in SELEX, the affinity of the final selected population is improved by three to five orders of magnitude relative to that of the starting random sequence population.

The first SELEX experiment was carried out against T4 DNA polymerase, a target that not only normally interacts with nucleic acid substrates, but also was known to be auto-regulated at the translational level by virtue of its ability to specifically recognize a 'translational operator' embedded within its own mRNA (Tuerk and Gold, 1990). The operator is a stem-loop with an eight-base loop. The library consisted of eight randomized positions and consisted of 4^8 or 65 536 different members. Flanking the random region were sequences capable of forming a stem. Thus, the goal of this experiment was to select the optimum loop sequence for the target. In this case, careful design of the binding conditions during SELEX rounds resulted in the definition of those primary sequences of eight residues with the highest affinity for the target present in the initial population. The outcome of the experiment was surprising. After only four rounds, only two sequences remained of the original population. The 'natural' sequence evolved to bind the protein was indeed identified as one 'winner' in the SELEX process. Curiously, a second sequence, not apparently related to the first in primary sequence, nor obviously predicted to be identical in conformation to the 'natural' site, showed identical affinity to the 'natural' site. Subsequent structural analysis by NMR has shown that these two sequences are in fact isomorphous (Mirmira and Tinoco, 1996).

Traditionally, therapeutics are small organic molecules discovered by screening libraries of molecules synthesized or isolated from natural sources. The notion

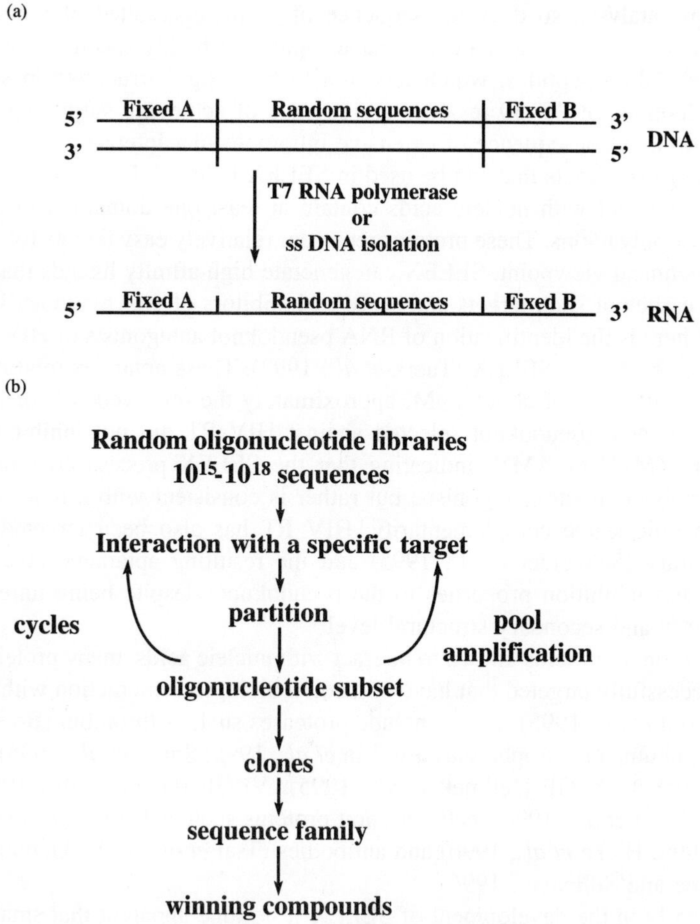

(a)

(b)

Fig. 1 Outline of the SELEX procedure. (a) Double-stranded DNA is depicted with two different fixed sequences flanking a region of random sequence. Transcription with T7 RNA polymerase yields RNA containing a portion of the Fixed A region, the random stretch, and the Fixed B region. This RNA is the material used in the first round of SELEX. (b) The experimental steps in SELEX.

that high affinity ligands could comprise nucleotides is a relatively new and somewhat iconoclastic idea. Its plausibility follows from an increasing appreciation of the complex sequence-dependent tertiary structures displayed by 'natural' oligonucleotides, such as tRNA, which was the first RNA whose structure was understood at high resolution (Kim *et al.*, 1974). Formation and stabilization of tertiary structure of an oligonucleotide occurs mainly through intramolecular Watson-Crick, non-Watson-Crick interactions and base stacking forces. The resulting structure can be far more elaborate and rich in terms of shape complexity than the double helix. Underlying SELEX is the idea that just as the amino acid sequence of a protein dictates a specific shape enabling recognition and often catalysis, so does the sequence of a single-stranded oligonucleotide determine a particular sophisticated shape capable of highly specific recognition. However, unlike peptides, which tend not to have rigid structures in solution, oligonucleotides of 20–30 residues are capable of achieving surprising rigidity in solution as a consequence of extensive intramolecular interactions.

The range of targets that can be used in SELEX is broad. Proteins with known ability to interact with nucleic acids contain at least one domain with intrinsic affinity for polyanions. These proteins represent relatively easy targets for SELEX from a technical viewpoint. SELEX can generate high affinity ligands that access the active sites of such targets and serve as inhibitors of enzymatic activity. An example here is the identification of RNA pseudoknot antagonists of HIV reverse transcriptase (RT) by SELEX (Tuerk et al., 1992). These aptamers inhibit cDNA synthesis with a Ki of about 2 nM, approximately the observed Kd for binding. Remarkably, the pseudoknot selected against HIV RT did not inhibit the RTs of either MMLV or AMV, indicating that the SELEX process does not yield generic polyanion site antagonists, but rather is consistent with a more intimate target-specific shape complementarity. HIV RT has also been targeted with a DNA library (Schneider et al., 1995) and the resulting aptamers have similar binding and inhibition properties to the pseudoknots despite being unrelated at the primary and secondary structural level.

In addition to proteins known to interact with nucleic acids, many proteins have been successfully targeted that have no known biological interaction with nucleic acids (Gold et al., 1995). These include proteases such as thrombin (Bock et al., 1992) and human neutrophil elastase (Lin et al., 1995; Smith et al., 1995) growth factors such as bFGF (Jellinek et al., 1995), VEGF (Green et al., 1995) and PDGF (Green et al., 1996), cell adhesion proteins such as L-selectin (O'Connell et al., 1996; Hicke et al., 1996) and antibodies (Tsai et al., 1992; Doudna et al., 1995; Lee and Sullenger, 1996).

Very early in the development of SELEX it became apparent that small molecules could also be interesting subjects of a SELEX approach. Ellington and Szostak (1990) showed that random RNA populations contained aptamers that specifically recognized organic dye molecules immobilized on a solid support. Peptides, such as the 9 amino acid peptide, substance P (Nieuwlandt et al., 1995) or a 17 amino acid alpha helical peptide derived from the HIV Rev protein (Xu and Ellington, 1996), have been targeted. A variety of aminoglycoside antibiotics, including tobramycin (Wang and Rando, 1995), neomycin (Wallis et al., 1995), lividomycin and kanamycin (Lato et al., 1995) have been targeted. Individual nucleoside triphosphates can be targets; an anti-ATP aptamer recognizes ATP with a Kd of 0.7 µM, and is capable of discriminating against dATP by more than three orders of magnitude (Sassanfar and Szostak, 1993). Aptamers specific for individual amino acids, such as L- and D-arginine (Famulock, 1994; Connel et al., 1993), valine (Majerfield and Yarus, 1994) and D-tryptophan (Famulok and Szostak, 1992) have also been described. Co-factors such as AND and cyanocobalomin have also yielded high affinity aptamers. At a more complex

level of target, aptamers have been generated to intact Rous sarcoma viral particles (Pan *et al.*, 1995). These aptamers, when stabilized to nuclease resistance,
were capable of blocking viral replication in cells challenged with RSV in tissue
culture (Pan *et al.*, 1995). These examples point to the robustness of the procedure and support the notion that its applicability is remarkably broad as a ligand
generation tool.

In this report I describe the basic procedure, some variations on it and a variety
of applications directed towards therapeutic and diagnostic targets.

2 Design and construction of an oligonucleotide library

The starting point for the library is a collection of single-stranded DNA molecules
with a random sequence of some length flanked by distinct fixed sequences
that serve as recognition sites for primers in PCR. This DNA is synthesized by
solid phase methodology on a commercial DNA synthesis machine programmed
to make the random region using a fifth precursor bottle containing all four
phosphoramidite precursors in close to equimolar concentrations. The precise
amounts of these precursors are actually adjusted for coupling efficiency differences so as to produce truly random sequences. It is important to characterize
the library for equimolar base composition before starting SELEX to ensure that
no biases in synthesis have occurred.

The length of the random region used by various investigators can range from
rather short stretches of 20–30 residues to very long regions of more than 200
bases. Some considerations in deciding the length of the random region have
involved speculations about the number of residues needed to comprise a high
affinity recognition site. One can look to 'natural' RNA-protein recognition sites
to gain some perspective on this issue. One of the best characterized RNA recognition sites is that of the bacteriophage R17 coat protein (Romaniuk *et al.*, 1987).
This protein specifically binds to a stem loop structure in which the loop contains
four bases of specific sequence (AUUA) and the stem is relatively generic,
comprised of four base pairs except for the required presence of a bulged A
residue. This small ensemble of bases specifically folded into a precise conformation binds to its target with a Kd of about 5 nM. A SELEX experiment with a
32 nt random region targeted to R17 coat protein yielded aptamers with this basic
motif except that the loop was AUCA which binds to the protein with slightly
better affinity than the natural site (Schneider *et al.*, 1992).

While it is clear that a high affinity site can be comprised of 15–25 residues,
what is the most efficient way to ensure the presence of such sites in the library?
Here some statistical and practical issues intersect. Sequence complexity for
oligonucleotide libraries scales as 4^n, where n is the number of random positions. With 30 random positions, a library of roughly 10^{18} different sequences is
possible. With 100 random positions, the number of possible sequences (about
10^{60}) approaches the number of elementary particles in the universe. Thus, from
a biochemical perspective, it is possible experimentally to explore the entire

sequence 'landscape' represented by a library of about 25–30 random positions, that is, determine definitively whether a high affinity aptamer exists in this library, while for a large library, only an infinitesimal fraction of the possible sequence and shape space can be experimentally searched. Synthesis of large random sequence libraries also involves some technical complexities, necessitating ligation of shorter stretches of sequences together to make up the starting library (Bartel and Szostak, 1993). Nonetheless, large random sequence libraries have been constructed and successfully utilized to find rare catalytic aptamers (see Ekland and co-workers (1995) for details on the design, properties and use of such libraries).

The decision about length of the random sequence used in a SELEX experiment can depend on the goal of the project and the method envisioned to produce the winning ligand. Typically, once a SELEX experiment is completed, a binding site is imbedded within the length of the once random region. This site may be composed of contiguous sequence but this is by no means necessarily the case. Motifs critical for binding may be interspersed with dispensable sequences. An important final goal of a SELEX experiment can be the definition of the minimal length high affinity site by a process termed truncation (see below). If a ligand of interest is to be used in a therapeutic application it must be relatively short because synthesis of aptamers at large scale (greater than 1 mmole) is at present not efficient at lengths greater than about 40 residues. For experimental uses where the aptamer can be produced by transcription, the need to truncate the winning ligand down to the smallest length retaining high affinity binding is not necessarily critical. Typically, random region length used in SELEX experiments at NeXstar have been in the 30–50 residue range. In several cases, SELEX experiments directed at the same target have been carried out with random regions of different length over a range of 30–60 residues. These experiments yielded similar aptameric families, indicating that at least over this range and for these targets, the optimal binders could be found in libraries composed of 30 random positions (N. Janjic, personal communication).

Once the initial DNA strand is synthesized, the second strand is made following annealing of a complementary primer at the $5'$ end, followed by extension using Klenow fragment of DNA polymerase I. The double-stranded DNA population can be transcribed into RNA by virtue of a promoter for T7 RNA polymerase present in the fixed sequence flanking the random region. Transcription to generate several nmoles of RNA produces the starting material for the first round of SELEX. The use of modified nucleoside triphosphates 'front-loaded' in SELEX has been extensively explored (Eaton and Pieken, 1995). Such modified NTPs offer two major advantages for therapeutic applications of SELEX. First, they confer nuclease resistance on the library. Second, they can expand the chemical repertoire of the library thus potentially increasing the possible interactions with residues on the target. The most commonly used modifications are $2'F$ or $2'$ NH_2 pyrimidine triphosphates following the observation that these modifications are sufficient to confer extensive stability to the library when incubated in human

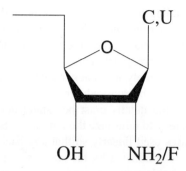

Fig. 2 Structure of the modified sugar residues on cytosine and uracil nucleotides used in SELEX to confer nuclease stability on the RNA population. Modifications are either $2'NH_2$ or $2'F$.

plasma (Fig. 2) (Pieken *et al.*, 1991). Unmodified RNA is degraded in seconds in human plasma while modified RNA is stable for many hours. Modifications on the bases have focused on the 5 position of uracil, directed away from the Watson-Crick pairing face, and thus are less likely to interfere with standard base-pairing interactions. Modifications used on the sugar or base moieties of nucleosides must be compatible with the enzymology used in SELEX i.e. T7 RNA polymerase, reverse transcriptase, and the enzymes in PCR. In general, many changes at the $2'$ position of the ribose and the 5 position of uracil have been found to be compatible with these enzymes (Eaton and Pieken, 1995).

3 Partition

Partition refers to the biochemical or physical process by which the fraction of the oligonucleotide library that has affinity for the target, and/or is capable of some selected function, is separated from the remaining members of the library. The partition process is extremely critical to a successful SELEX experiment. If it is inefficient or suffers from extensive background problems due to non-specific binding, SELEX progress can be slow. The efficiency of the partition process at each round determines the number of rounds required to reduce the library sequence complexity from a starting point of about 10^{14} sequences down to a manageable number, say 100 different sequences each of which can be identified and studied after cloning. Partition systems range from simple processes such as filtration through nitrocellulose filters, to chromatographic fractionation in which the target is immobilized on a solid support, to gel electrophoretic partition in which target-bound oligonucleotides are physically separated from unbound oligonucleotides. Nitrocellulose partition is fast, convenient and widely used in SELEX. Care must be taken to ensure that the free target is quantitatively retained on the filter and that during the course of rounds, background binding to nitrocellulose itself is controlled. Immobilization of the target on a solid support such as a column resin or a bead has numerous advantages including the ability to control background binding by extensive washing, and controlling the nature of the interaction of bound oligonucleotides to the target by specific elution schemes. This approach is especially attractive when the target is a small molecule and other partition schemes are precluded. Gel electrophoretic separation can be attractive because of extremely low backgrounds when the retarded labelled oligonucleotides are detected by autoradiography. This method is, of course, useful for protein targets and details must be established in pilot experiments. Overall, a productive strategy to minimize the number of SELEX rounds involves multiple partition processes used linearly so that any background binding that is observed in one process is eliminated by the successive partition method.

An example of how partition can be used to drive specificity of binding of oligonucleotides to a target is provided by SELEX experiments directed against theophylline, a xanthine base used in the therapeutic treatment of asthma (Jenison et al., 1994). Due to its narrow therapeutic index, serum levels of theophylline must be carefully monitored diagnostically. Because theophylline and caffeine differ only by the presence of a proton at N7 in theophylline compared to a methyl group in caffeine, cross recognition is an issue. The SELEX process used is outlined in Fig. 3. The partition process used involved covalent immobilization of theophylline via the C1 position to Sepharose. The oligonucleotide library was passed over the column and extensively washed. To eliminate those oligonucleotides with affinity for both theophylline and caffeine, the column was then treated with a high excess of free caffeine and further washed. Finally, free excess theophylline was added to the column and oligonucleotides eluted were isolated, reverse transcribed and amplified for the next round of SELEX. Thus, oligonucleotides whose binding specifically depended on discrimination between two very similar targets were selected. The oligonucleotides that emerged after the seventh round of SELEX had an apparent Kd of about 100 nM for theophylline and bound to caffeine about 10 000 fold less tightly (Jenison et al., 1994). This discrimination is about 100-fold more efficient than that reported for antibodies (Poncelet et al., 1990).

Progress during successive SELEX rounds is typically monitored by measuring affinity of the pools for the target compared to the initial random pool. When additional rounds fail to show increased affinity, and a bulk sequencing experiment indicates that the pool is no longer random, SELEX is done. The members of the selected pool are then cloned into a bacterial plasmid vector and individual members then sequenced. Typically, families of related sequences emerge from SELEX displaying a 'pseudo'-phylogeny that indicates the primary sequence motifs responsible for specific target recognition (Fig. 4). Sequence covariation among family members provides critical clues in this analysis. An important exercise is to organize the sequences into related families, then attempt to deconvolute the families into unique secondary structures. Sequence alignment is usually not difficult. For most SELEX experiments, it is possible to make specific predictions about the key secondary and often even tertiary structures responsible for binding, simply by analysis of the sequence families at the primary sequence

(a)

Theophylline   Caffeine

(b)

Theophylline column

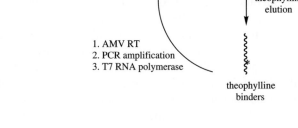

(c)

radiolabelled
RNA mixture

theophylline
elution

1. AMV RT
2. PCR amplification
3. T7 RNA polymerase

theophylline
binders

(d)   'Counter-SELEX' imparts Specificity

1. Caffeine
elution

2. Theophylline
elution

theophylline
binders that
do not bind
caffeine

Fig. 3 Outline of SELEX carried out against theophylline. (a) Structures of theophylline and caffeine. (b) Immobilization of theophylline via the C1 position on a solid support (dark ball). (c) SELEX rounds were carried out using excess soluble theophylline as a specific eluant. (d) A variation on the elution scheme used excess soluble caffeine to elute aptamers capable of binding caffeine. These are discarded. Aptamers resistant to caffeine challenge are eluted by theophylline.

Sequence Alignment	Secondary Structure	Target

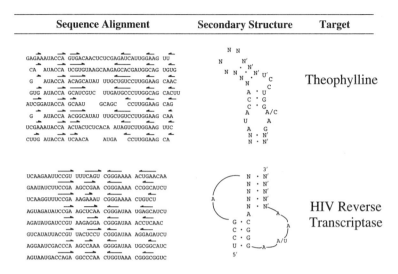

Fig. 4 Examples of SELEX-derived aptamer families at the primary and secondary structure levels. At the left are aligned sequences derived from SELEX carried out against either theophylline or HIV reverse transcriptase. Arrows above the individual sequences depict regions of intramolecular complementarity. The secondary structure shown represents a consensus derived from inspection of each sequence. N represents any nucleotide; N–N', any canonical base pair.

level. Software to facilitate alignment of SELEX derived sequences has been recently described (Davis *et al.*, 1996). Competition binding analysis can be done to determine whether different aptamers recognize identical or overlapping regions of the target. It is commonly observed that after extensive selection, one epitope of the target, presumably with the highest intrinsic affinity for oligo-nucleotides, predominates in the SELEX process. However, for certain targets, such as thrombin, which contains two potential anion binding sites, it is possible to find non-competing families of aptamers at the termination of the SELEX process. The DNA aptamer selected to thrombin binds in the fibrinogen binding exosite (Paborsky *et al.*, 1993). The RNA winner, in contrast, appears to bind in the heparin binding site (Kubik *et al.*, 1994).

4 Truncation

Truncation refers to the process by which the minimum length aptamer containing the high affinity target recognition site is identified. If the motif is obvious from primary sequence analysis, 'rational' truncates can be prepared and tested for affinity. For example, all members of a family may contain a predicted stem-loop secondary structure which co-varies at particular positions. A consensus sequence is determined, synthesized and tested for affinity to determine the validity of the assignments. Systematic analysis of the minimal binding site requires additional experimentation. In general, the approach is to digest a defined end-labelled aptamer from the non-labelled terminus in such a way as to generate all possible lengths, then to partition this mixture against the target under stringent binding conditions, analysing the length of the oligonucleotides that retain high affinity binding ability. Repeating this process from the opposite terminus provides the other boundary. The result of this analysis yields the shortest contiguous length required for target recognition. Although it is difficult to generalize about length requirements for high affinity binding, it is often observed for protein and small molecular weight targets that truncates in the range of 20–40 residues form structures that confer specific recognition of the target. These lengths are consistent with those of 'natural' RNA targets discussed earlier.

5 Post-SELEX modifications

Following the identification of a truncated aptamer with high affinity for the target, it has been observed that the further substitution of modified nucleotides for 2'OH ribopurine nucleotides can result in enhanced stability and affinity when done selectively. For example, an anti-VEGF aptamer initially isolated from a 2'-NH2 pyrimidine library contained 14 ribopurine residues (Green et al., 1995). If all of these residues were substituted with 2'-O-methylpurines, binding affinity for VEGF was reduced by a factor of 100, consistent with the notion that specific structural motifs are dependent on one or more 2'OH ribose residues. However, substitution of 10 of the 14 residues with 2'-O-methylpurines improved the affinity of the initial aptamer 10-fold and improved the stability of the aptamer 8-fold to nuclease degradation (Green et al., 1995).

6 Aptamer structures

Our understanding of RNA tertiary structure is limited by the small number of examples that have been solved at high resolution. Aptamers represent attractive models with which to expand this database and begin to provide rules by which 'the RNA folding problem' can be approached. The supposition that aptamers identified by SELEX would possess sophisticated tertiary conformations permitting specific interactions with targets has been validated by the initial wave of high resolution structure analysis of aptamer-target complexes (see Feigon and co-workers (1996) for a review). The first aptamer for which detailed solution structure was determined was a 15-nucleotide DNA aptamer that binds to and inhibits thrombin (Bock et al., 1992). This aptamer folds into a unimolecular structure containing two G-quartet motifs (Schultze et al., 1994). The X-ray crystal structure of the thrombin aptamer complex has been reported (Padmanabhan et al., 1993); the aptamer is sandwiched between the anionic fibrinogen binding site of one thrombin and the heparin binding site of another. Aptamers to small molecules have been especially valuable as structural models. The structure of an RNA aptamer–AMP complex has recently been solved using multidimensional NMR techniques by two groups (Dieckmann et al., 1996; Jiang et al.,

1996). This aptamer, initially isolated by virtue of its ability to bind to immo-bilized ATP, also recognizes ADP and AMP with equal affinity (Sassanfar and Szostak, 1993). The AMP binding pocket is, in secondary structural terms, an 11-nucleotide internal loop opposite a G bulge, flanked by non-conserved stems. However, the loop actually exists as a complex, compact and highly folded struc-ture that envelops the AMP molecule. This structure contains two G–G pairs and a radical change in backbone direction called a U-turn. These initial struc-tures emphasize the remarkable plasticity of the phosphodiester backbone and the unanticipated ability of non-canonical nucleotide pairs and triples to create high-specificity binding pockets for targets. Additional aptamer structures will emerge over the next several years and further expand these possibilities.

7 Applications

The first aptamer tested for function *in vivo* was directed against thrombin. This 15mer DNA aptamer forms a G-quartet structure in solution (Padmanabhan *et al*.,1993; Macaya *et al*., 1993) and is capable of prolonging blood clot time *in vitro* (Bock *et al*., 1992). The DNA aptamer also prolongs the prothrombin time after infusion into cynomolgous monkeys in a dose dependent manner, and has been shown to retard clot formation in an *ex vivo* whole artery angioplasty model system (Li *et al*., 1994; Griffin *et al*., 1993). Significantly, the aptamer is capable of inhibition of clot-bound thrombin in contrast to other agents such as heparin (Li *et al*., 1994).

Recently, aptamers to antibodies related to those present in patients with the autoimmune disease myasthenia gravis have been isolated (Lee and Sullenger, 1997). These antibodies recognize the acetylcholine receptor and are thought to be implicated in the etiology of this disorder. The aptamers were generated from a 2′-aminopyrimidine library and specifically recognized antibodies present in the serum of patients with myasthenia gravis. Antagonists to the collection of anti-cholinesterase antibodies present in these patients is a potential novel therapeutic approach.

A third example of aptamer utility *in vivo* is provided by efforts to develop antagonists of cell surface adhesion proteins in the selectin family. The selectins are a family of calcium-dependent cell surface lectins that mediate cell adhesion through recognition of cell specific carbohydrate ligands. Three major members are known, designated E, L and P selectin. Selectin-ligand recognition is known to play a role in leukocyte extravasation from the vasculature into tissues in response to signals associated with inflammation (Lasky, 1995; McEver *et al*., 1995). L-selectin is constitutively expressed on the surface of most leukocytes, while E- and P-selectin are expressed inducibly on endothelial cells and/or platelets (Springer, 1994). Because selectins have been implicated in reperfusion injury following ischaemia, we have carried out a number of separate SELEX experi-ments targeting L-selectin with a variety of modified ribose libraries, including $2'NH_2$ pyrimidines, $2'F$ pyrimidines, as well as a DNA library, to isolate aptamers that might interfere with selectin function. These aptamers have been compared for performance in binding to purified L-selectin in solution and on leukocytes (Hicke et al., 1996; O'Connel et al., 1996).

To facilitate partitioning, L selectin was expressed as a fusion protein in mammalian cells. The protein was immobilized on Protein A Sepharose beads and exposed to the SELEX library. To isolate aptamers that recognized the lectin binding domain, elution of bound oligonucleotides was carried out with low concentrations of EDTA, which was intended to disrupt Ca^{++}-dependent interactions between library members and the target. The affinity of the starting $2'$ NH_2 pyrimidine pool for the bead-bound protein was $9 \mu M$; after 14 rounds of selection the pool had a Kd of $0.3 \, nM$, representing a 300 000-fold improvement in affinity over the starting pool. The interaction of the pool with L-selectin was divalent cation dependent, consistent with targeting of the lectin domain. Material from the tenth round was tested in a static ELISA assay to determine if it blocked the interaction of L-selectin with its major carbohydrate ligand, sialyl Lewis[x], immobilized on a plastic surface. This material blocked the interaction with IC_{50} of $18 \, nM$, demonstrating directly that aptamers bound at or near the lectin recognition domain. It is especially significant that this pool had no effect on the interaction of P-selectin or E-selectin with immobilized sialyl Lewis[x], indicating that the interaction is specific for the L-selectin lectin domain.

These observations provided a context in which to test the idea that aptamers directed against a cell surface receptor could interfere with receptor function *in vivo*. To carry out this work three issues had to be addressed. First, the L-selectin aptamer had to interact with its target at physiological temperatures. Second, the aptamer had to be formulated to be retained in the vasculature for a sufficient time to be efficacious. Third, a human L-selectin dependent model had to be developed for use in an animal. It was observed that the initial set of aptamers containing $2'NH_2$ pyrimidine modifications bound to L-selectin in a temperature dependent manner such that high affinity was observed at $4°C$ and $22°C$, but not at $37°C$, precluding the use of these aptamers *in vivo*. The reason for this behaviour is unknown but could be related to the lowered thermal stability of $2' \, NH_2$ helices relative to $2'F$ or $2'H$ containing helices (Aurup et al., 1994). Consequently, SELEX was repeated using a DNA library and incorporating a requirement for binding at $37°C$. This SELEX yielded DNA aptamers with a Kd of about $1 \, nM$ for the pool after 15 rounds (Hicke et al., 1996). A specific aptamer in this pool, called LD201, was truncated to a length of 49 residues, and shown to inhibit L-selectin interaction with sialyl Lewis[x] with an IC_{50} of about $3 \, nM$ at $37°C$. After conjugation to FITC this aptamer was shown to bind to human lymphocytes and neutrophils in whole blood by flow cytometry. This binding was saturable, divalent cation dependent, and could be blocked by addition of a monoclonal antibody specific for L-selectin (Hicke et al., 1996).

A key property of any new therapeutic modality is appropriate pharmacokinetic behaviour. Aptamers are large relative to traditional drugs and are polyanions. We have observed that nuclease resistant SELEX derived aptamers are rapidly cleared from circulation after injection intravenously, probably due to their small size (S. Gill, unpublished). However, addition of a 20 000 mol wt polyethylene glycol tail to the 5' terminus of aptamers can be shown to greatly prolong the circulation of the aptamer after intravenous injection in rats or mice (S. Gill, unpublished). The PEG is coupled via a succinamide ester to the 5' terminus of the aptamer which contains an amine moiety. The 20 000 mol wt PEG formulation is one of several that have been explored in the model described below and represents a start towards a detailed understanding of the behaviour of aptamers *in vivo*.

Aptamers selected to recognize human L-selectin bound to human but not rodent lymphocytes, indicating their high degree of specificity. To assess their ability to block an L-selectin dependent process *in vivo*, this specificity necessitated a xenogeneic approach. In this model, human PBMC, labelled with ^{51}Cr, were injected intravenously into SCID mice. The human cells migrate to peripheral and mesenteric lymph nodes in an L-selectin dependent manner, as determined by the observation that this distribution is blocked by DREG-56, a monoclonal antibody specific for human L-selectin, but not by MEL-14, a monoclonal antibody specific for murine L-selectin (Hicke *et al.*, 1996). In initial experiments, aptamers were mixed with the human cells prior to their injection into SCID mice and the L-selectin aptamer, but not a scrambled sequence control aptamer, was shown to block trafficking to PLN and MLN in a dose-dependent manner. A more rigorous assessment of performance was to introduce the aptamer 1–5 minutes prior to the cells. Under these conditions, at concentrations of 4 nmol/mouse, specific inhibition of trafficking by the L-selectin aptamer was observed (Hicke *et al.*, 1996).

These observations have several implications for aptamers as potential therapeutics. First, they can be extremely specific for their target. Second, they can be selected to a purified molecular species in a physiological buffer, then shown to recognize the same target in the context of a cell surface. Third, the aptamer can be simply formulated to improve its retention in serum without compromising its specificity. Fourth, at low doses, the formulated aptamer can interfere with normal functions of the target in whole blood, suggesting that potential interference with other components of blood is not significant with regard to function.

8 Conclusion and perspective

Aptamers represent a new combinatorial chemistry approach to drug discovery. Although superficially similar to antisense compounds, they differ profoundly from them in that their targets are not nucleic acids. Indeed, potential targets for aptamers are unlimited, ranging from small organic molecules to entire cell surfaces. It is perhaps more useful to compare SELEX-derived aptamers to antibodies. From the perspective of binding performance, aptamers can show affinities and specificities comparable or better than antibodies. The ability to chemically synthesize and modify aptamers offers novel advantages in regard to formulation. Monomeric aptamers can be multimerized to gain improvement in target binding through reduction in effective off-rate. They can be incorporated into liposomes to alter their pharmacokinetic and biodistribution profile.

Attempts to raise antibodies to aptamers have indicated that they are not immunogenic in rats (Biesecker and Bendele, unpublished). Although they are large, complicated molecules, methods for their economic manufacture at scale are rapidly emerging.

The process of evaluating aptamers as therapeutic agents in animal models of disease has just begun. There is much to be learned about their behaviour *in vivo*. The next few years will witness whether the surprising performance of aptamers as reagents *in vitro* and *ex vivo* can be translated into efficacy in the clinic.

Acknowledgements

I thank my colleagues at NeXstar for insight and critical comments. Special thanks to David Parma, Susan Watson, Brian Hicke, Stan Gill, Ray Bendele, Nebojsa Janjic, Tim Fitzwater, Rob Jenison, David Sebesta, Dom Zichi and Larry Gold for input, and Amy Crossman for help in preparing the manuscript.

References

Aurup, H., Tuschi, T., Benseler, F., Ludwig, J. and Eckstein F. (1994). *Nucleic Acids Res.* **22**, 20–24.
Bartel, D. P. and Szostak, J. W. (1993). *Science* **261**, 1411–1418.
Bock, L. C., Griffin, L. C., Latham, J. A., Vermaas, E. H. and Toole, J. J. (1992). *Nature* **355**, 564–566.
Connell, G. J., Illangesekare, M. and Yarus, M. (1993). *Biochemistry* **32**, 5497–5502.
Davis, J. P., Janjic, N., Javornik, B. E. and Zichi, D. (1996). *Methods Enzymol.* **267**, 302–314.
Dieckmann, T., Suzuki, E., Nakamura, G. K. and Feigon, J. (1996). *RNA* **2**, 628–640.
Doudna, J. A, Cech T. R. and Sullenger, B. A. (1995). *Proc. Natl. Acad. Sci. USA* **92**, 2355–2359.
Eaton, B. and Pieken, W. (1995). *Annu. Rev. Biochem.* **64**, 837–863.
Ekland, E. H., Szostak, J. W. and Bartel, D. P. (1995). *Science* **269**, 364–370.
Ellington, A. D. and Szostak, J. W. (1990). *Nature* **346**, 818–822.
Famulok, F. M. (1994). *J. Am. Chem. Soc.* **116**, 1698–1706.
Famulok, M. and Szostak, J. (1992). *J. Am. Chem. Soc.* **114**, 3990–3991.
Feigon, J., Dieckmann, T. and Smith, F. W. (1996). *Chem. Biol.* **3**, 611–617.
Gold, L., Polisky, B., Uhlenbeck, O. and Yarus, M. (1995). *Annu. Rev. Biochem.* **64**, 763–797.
Green, L. S., Jellinek, D., Bell, C., Beebe, L., Feistner, B. D., Gill, S. C., Jucker, F. M. and Janjic, N. (1995). *Chem. Biol.* **2**, 683–695.
Green, L. S., Jellinek, D., Jenison, R., Ostman, A., Heldin, C-H. and Janjic, N. (1996). *Biochemistry* **35**, 14413–14424.
Griffin, L. C., Tidmarsh, G. F., Bock, L. C., Toole, J. J. and Leung, L. L. K. (1993). *Blood* **81**, 3271–3276.
Hicke, B. J., Watson, S. R., Koenig, A., Lynott, K. C., Bargatze, R. F., Chang, Y-F., Ringquist, S., Moon-McDermott, L., Jennings, S., Fitzwater, T., Han, H-L., Varki, N.,

176 **B. Polisky**

Albinana, I., Willis, M. C., Varki. A. and Parma, D. (1996). *J. Clin. Invest.* **98**, 2688-2692.
Jellinek, D., Green, L. S., Bell, C., Lynott, C. K., Gill, N., Vargeese, C., Kirschenheuter, G., McGee, D. P. C., Abesinghe, P., Pieken, W. A., Shapiro, R., Rifkin, D. B., Moscatelli, D. and Janjic, N. (1995). Biochemistry **34**, 11363-11372.
Jenison, R. D., Gill, S. C., Pardi, A. and Polisky, B. (1994). *Science* **263**, 1425-1429.
Jiang, F., Kumar, R. A., Jones, R. A. and Patel, D. J. (1996). *Nature* **382**, 183-186.
Kim, S. H., Suddath, F. L., Quigley, G. J., McPherson, A., Sussman, J. L., Wang, A. H., Seeman, N. C. and Rich, A. (1974). *Science* **185**, 435-440.
Kubik, M., Stephens, A. W., Schneider, D. A., Marlar, R. and Tassett, D. (1994). *Nucleic Acids Res.* **22**, 2619-2626.
Lasky, L. A. (1995). *Ann. Rev. Biochem.* **64**, 113-139.
Lato, S. M., Boles, A. R. and Ellington, A. D. (1995). *Chem. Biol.* **2**, 291-303.
Lauhon, C. T. and Szostak, J. W. (1995). *J. Am. Chem. Soc.* **117**, 1246-1257.
Lee, S-W. and Sullenger, B. A. (1996). *J. Exp. Med.* **184**, 315-324.
Lee, S-W. and Sullenger, B. A. (1997). *Nature Biotechnology* **15**, 41-45
Li, W-X., Kaplan, A. V., Grant, G. W., Toole, J. J. and Lueng, L. L. K. (1994). *Blood* **83**, 677-682.
Lin, Y., Padmapriya, A., Morden, K. M. and Jayasena, D. (1995). *Proc. Natl. Acad. Sci. USA* **92**, 11044-11048.
Macaya, R. F., Schultze, P., Smith, F. W., Roe, J. A. and Feigon, J. (1993). *Proc. Natl. Acad. Sci. USA* **90**, 3745-3749.
Majerfield, I. and Yarus, M. (1994). *Nature Struct. Biol.* **1**, 287-292.
McEver, R. P., Moore, K. L. and Cummings, R. D. (1995). *J. Biol. Chem.* **270**, 11025-11028.
McGee, G., Abesinghe, D. P. C., Pieken, W. A, Shapiro, R., Rifkin, D. B., Moscatelli, D. and Janjic, N. (1995). *Biochemistry* **34**, 11363-11372.
Mirmira, S. R. and Tinoco I. (1996). *Biochemistry* **35**, 7675-7683.
Nieuwlandt, D.,Wecker, M. and Gold, L. (1995). *Biochemistry* **34**, 5651-5659.
O'Connell, D., Koenig, A., Jennings, S., Hicke, B., Han, H-L., Fitzwater, T., Chang, Y-F., Varki, N., Parma, D. and Varki, A. (1996). *Proc. Natl. Acad. Sci. USA* **93**, 5883-5887.
Paborsky, L. R., McCurdy, S. N., Griffin, L. C., Toole, J. J. and Leung, L. L. K. (1993). *J. Biol. Chem.* **268**, 20808-20811.
Padmanabhan, K., Padmanabhan, K. P., Ferrara, J. D., Sadler, J. E. and Tulinsky, A. (1993). *Proc. Natl. Acad. Sci. USA* **268**, 17651-17654.
Pan, W., Craven, R. C., Qiu, Q., Wilson, C. B., Willis, J. W., Golovine, S. and Wang, J. (1995). *Proc. Natl. Acad. Sci. USA* **92**, 11509-11513.
Pieken, W. A., Olsen, D. B., Benseler, F., Aurup, H. and Eckstein, F. (1991). *Science* **253**, 314-317.
Poncelet, S. M., Limet, J., Noel, J. P., Kayaert, M. C., Galanti, L. and Collet-Cassart, D. (1990). *J. Immunoassay* **11**, 77-88.
Romaniuk, P., Lowary, P., Wu, H-N., Stormo, G. and Uhlenbeck, O. (1987). *Biochemistry* **26**, 1563-1568.
Sassanfar, M. and Szostak, J. (1993). *Nature* **364**, 550-553.
Schneider, D., Tuerk, C. and Gold, L. (1992). *J. Mol. Biol.* **228**, 862-869.
Schneider, D. J., Feigon, J., Hostomsky, Z. and Gold, L. (1995). *Biochemistry* **34**, 9599-9610.
Schultze, P., Macaya, R. F. and Feigon, J. (1994). *J. Mol. Biol.* **235**, 1532-1547.
Seeman, N. C. and Rich, A. (1974). *Science* **185**, 435-440.
Smith, D., Kirschenheuter, G. P., Charlton, J., Guidot, D. M., Repine, J. E. (1995). *Chem. Biol.* **2**, 41-750.
Springer, T. A. (1994). *Cell* **76**, 301-314.

Tsai, D. E., Kenan, D. J. and Keene, J. D. (1992). *Proc. Natl. Acad. Sci. USA* **89**, 8864–8868.
Tuerk, C. and Gold, L. (1990). *Science* **249**, 505–510.
Tuerk, C., MacDougal, S. and Gold, L. (1992). *Proc. Natl. Acad. Sci. USA* **89**, 6988–6992.
Wallis, M. G., Von Ashen, U., Schroeder, R. and Famulok, M. (1995). *Chem. Biol.* **2**, 543–552.
Wang, Y. and Rando, R. R. (1995). *Chem. Biol.* **2**, 281–290.
Xu, W. and Ellington, A. D. (1996). *Proc. Natl. Acad. Sci. USA* **93**, 7475–7480.

12

Catalysis of Organic Reactions by RNA — Strategies for the Selection of Catalytic RNAs

ANDRES JÄSCHKE

Institut für Biochemie, Freie Universität Berlin, Thielallee 63, D-14195 Berlin, Germany

Abstract

SELEX is a powerful methodology for the identification of RNA or DNA molecules with specific binding properties or catalytic activities. Species satisfying a selection criterion are separated from the vast excess of unfunctional sequences in a partitioning step and then enzymatically amplified. The most successful technique for the isolation of catalytically active species among the huge number of random sequences is direct selection. In this process, the catalytic properties of the desired molecule can be directly exploited for their iterative enrichment. RNA pool molecules are incubated with a potential substrate and those RNA molecules are selected that form a covalent bond with the substrate and thereby acquire a new characteristic property allowing separation from unreacted species. While the use of direct selection has led to the isolation of several new ribozymes, it has some inherent drawbacks limiting its broader application. The most serious limitation is that one of the reactants is always the RNA itself, therefore only allowing the isolation of catalysts for self-modication reactions of RNA. Catalysts for reactions between small reactants, being the basis of both biochemistry and organic synthesis, cannot be isolated using this strategy.

SELEX with linker-coupled reactants is an improvement of direct selection designed to expand the chemistry of RNA catalysis to bimolecular reactions involving two small non-oligonucleotide substrates. Thereby, one of the potential reactants is site-specifically attached to the RNA pool molecules via a long flexible polymeric linker, while the other reactant carries an affinity tag, such as biotin. After reaction, the products are linked to both the RNA and the affinity tag,

THE MANY FACES OF RNA
ISBN 0-12-233210-5

allowing isolation and amplification of functional species. The incorporation of a cleavage site in the linker gives additional control of the reaction site. SELEX schemes have been developed for the selection of catalysts for Diels-Alder reactions and for nucleophilic substitution reactions. The application of similar selection schemes to bond cleavage and redox reactions will contribute to the exploration of the potential and limits of RNA catalysis, and applications in diagnostics and drug development as well as in organic synthesis are conceivable.

1 Introduction

Combinatorial RNA libraries have found increasing use for the identification of new ligands and catalysts (Gold *et al*., 1995; Lorsch and Szostak, 1996). Commercial interest has been generated by RNA ligands with highly specific binding properties, so-called aptamers, and by the ability of ribozymes to sequence-specifically cleave RNA molecules. The two previous chapters gave a number of examples illustrating the potential of RNA-derived molecules as drugs and diagnostics.

Since the discovery of enzymatically active RNA just over a decade ago, tremendous interest has grown in the isolation of new ribozyme activities, not only to explore the catalytic potential and the limits of RNA catalysis but also to test the hypothesis that, at some time in the past, life was based on RNA. This question provided much of the impetus in the development of *in vitro* RNA selection (Williams and Bartel, 1996). With the ability to isolate rare functional individuals from vast pools of sequence variants, molecules have been generated that dramatically extend the known range of RNA binding and catalytic activities.

The general methodology of *in vitro* selection (Ellington and Szostak, 1990; Tuerk and Gold, 1990) has been described in the previous chapter. A pool of RNA sequence variants is subjected to a selection (or partitioning) step, and the selected molecules are copied and enzymatically amplified. The ability to iterate the selection–amplification cycle ideally allows the isolation of the most active sequences, even when they are exceedingly rare in the initial pool or when they have only a small advantage over competitors. This iterative procedure is possible only because nucleic acids can be copied enzymatically, and is not available for organic small-molecule libraries and chemically synthesized peptide or carbohydrate libraries. Due to this property, recursive deconvolution, tagging or binary coding is unnecessary, and the complexity of libraries used in SELEX experiments is huge compared to other classes of compounds.

Selection has been applied to RNA species that (i) bind a target molecule (Jenison *et al*., 1994), (ii) serve as a substrate for an enzyme (Liu and Altman, 1994), or (iii) perform a self-modifying reaction (Bartel and Szostak, 1993), and categories (i) and (iii) are relevant for the isolation of RNA catalysts.

This chapter focuses on the selection of RNA molecules with catalytic properties. First, the advantages and limitations of currently used strategies for the selection of catalytic RNAs will be discussed, then some recent work from the author's laboratory will be introduced aimed to overcome these limitations. Alternative strategies to select for catalytic activity will be proposed, and at the end some ideas about implications and future applications of RNA catalysis will be presented.

2 Selection of catalytic RNA molecules

There are currently two strategies for the selection of catalytic RNA molecules, namely selection against transition state analogues (TSAs) and direct selection. The principle of the selection against TSAs goes back to the ideas of Haldane and Pauling that any substance that stabilizes a transition state is a catalyst of the respective reaction (Haldane, 1930; Pauling, 1946). Within the last years, numerous catalytic antibodies have been generated by challenging the immune system with transition state analogues (Lerner *et al.*, 1991), and the success of this strategy prompted several laboratories to fish for catalytic RNAs in a two-step process. In the first step, a chemically synthesized stable analogue mimicking the geometry and charge distribution of the transition state of the reaction under investigation is immobilized on a matrix and used to select RNA aptamers that bind to these transition state analogues using standard SELEX protocols. In the second step, the selected RNAs are then screened for their ability to catalyse the respective reaction (Prudent and Schultz, 1996).

While the first part, the identification of aptamers for the TSAs is generally unproblematic, these only occasionally exhibit any catalytic activity. Currently, there are only two types of reactions where this selection strategy has proven successful; an RNA catalyst for the isomerization of a bridged biphenyl system has been described (Prudent *et al.*, 1994), and the metallation of porphyrins has been achieved by two different groups (Conn *et al.*, 1996; Li and Sen, 1996). These success stories, however, are by far outnumbered by failed attempts in several different laboratories (Morris *et al.*, 1994; Prudent and Schultz, 1996).

The most successful strategy, however, is direct selection. Direct selection requires that the desired reaction be configured so that the catalytic molecules are themselves changed in a way that provides a basis for their enrichment during the selection step (Williams and Bartel, 1996). Selection criterion is ideally the catalysis of a chemical reaction. However, as a catalyst is per definition strictly required to leave the reaction unchanged, catalysis must be coupled with some other principle that leads to a selectable change in the catalyst molecule, and the way this is achieved is self-modification, letting one and the same molecule act as both catalyst and substrate. The principle is illustrated in Fig. 1. An RNA pool is incubated with a potential reactant that carries an anchor group, which is either a specific functional group (e.g. thiol, thiophosphate, primary aliphatic amine), or an affinity tag (e.g. biotin). Some RNA molecules react and are then covalently linked to that anchor group. These species can be specifically separated from the unreacted excess using suitably derivatized solid supports (e.g. activated thiopropyl agarose or streptavidin agarose). The isolated species are then enzymatically amplified, and the scheme is repeated several times. Other

Reaction: RNA + X ⟶ RNA-X

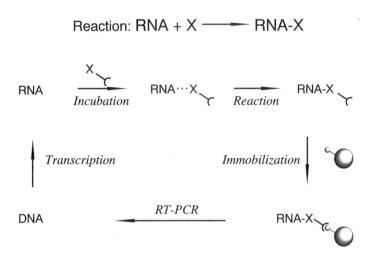

Fig. 1 Direct selection of self-modifying RNAs.

possibilities are that on catalysing a self-modification reaction, RNA molecules shorten or lengthen themselves and can be isolated by their altered electrophoretic mobility. For example, RNA molecules were selected that catalyse the transfer of the γ-phosphate of ATP to the 5'-terminal hydroxyl group of RNA by using a thiophosphate analogue of ATP during the selection (Lorsch and Szostak, 1994). After incubation, a thiophosphate was attached to some RNA molecules and these could be selectively isolated by using activated thiopropyl sepharose. Unmodified RNA was washed away, the disulfide bridge cleaved, and the selected RNA molecules enzymatically amplified. Similar schemes have been successfully applied to the identification of RNA catalysts for ligation, alkylation, acylation and transesterification reactions (Bartel and Szostak, 1993; Illangasekare *et al.*, 1995; Wilson and Szostak, 1995). In fact, direct selection of RNA catalysts is also a two-step procedure, since performing a self-modifying reaction is not identical with catalysis. In order to satisfy the selection criterion it is sufficient to carry out the reaction just once, and catalytic turnover is not in the selection query. The second step in identifying new catalysts is normally dissecting the self-modifying RNAs into a large catalyst part and a small substrate molecule and then studying the intermolecular action of the catalyst on the substrate (Lorsch and Szostak, 1994). This 'ribozyme engineering' was found to work well in a number of cases, however with comparatively low turnover numbers. This low turnover is presumably caused by the fact that substrate–catalyst interactions are largely mediated by base-pairing interactions.

Despite the success of direct selection, there are several drawbacks that severely limit its application:

(i)   In order to be selected, the RNA molecules must possess both substrate and catalyst properties. The whole selection principle relies on the catalyst acting *in cis* (that is, in an intramolecular reaction). If catalysis occurs *in trans* – intermolecular – with one pool RNA molecule being the catalyst and another being the substrate, then after the reaction only the former substrate molecule will carry the anchor group. A highly efficient catalyst in the pool having no substrate properties at all will be lost during election.

(ii) In all direct selection schemes published so far, one of the reactants was the RNA itself. Direct selection is an excellent tool to find catalysts for self-modification reactions of RNA, and the first successful attempts to expand such chemistry beyond the few functional groups occurring in nucleic acids have been published recently, utilizing modified nucleotides incorporated into RNA (Jensen *et al.*, 1995; Wecker *et al.*, 1996) or terminally modified complementary oligonucleotides (Smith *et al.*, 1995). The selected RNAs, however, display their catalytic activity only towards substrates presented in the context of an RNA molecule or oligonucleotide. It has so far been impossible to select catalysts for reactions between two small non-oligonucleotide substrates. Contemporary biochemistry, organic synthesis and prebiotic chemistry are all based on reactions between small substrates.

(iii) The third general limitation is that the selection criterion is just the attachment of the anchor group to the RNA, no matter where this occurs, no matter with which stereochemistry or by what kind of reaction mechanism. As long as the products can be enzymatically amplified, they are carried through the SELEX process.

3 SELEX with linker-coupled reactants

To overcome the limitations described above, we have devised a direct selection methodology using linker-coupled reactants. The general scheme is illustrated in Fig. 2. To find a catalyst for a general bimolecular reaction (A + B → AB), a long flexible polymeric tether is used to covalently link one of the potential reactants (in this case reactant A) to the pool RNA. Reactant B carries the anchor group (e.g. biotin), and after reaction, the products AB would be linked to both the anchor group (allowing isolation) and the RNA (allowing amplification). These molecules are isolated, amplified by reverse transcription and PCR, and the next SELEX round starts with the transcription and incorporation of linker-coupled reactant A. The whole scheme is not too different from the typical direct selection scheme and, in fact, this is again self-modification, but of complex RNA conjugate libraries rather than unmodified RNA pools. However, the attachment of a linker-coupled reactant should allow the expansion of RNA catalysis beyond self-modification, ideally to any bimolecular reaction that could happen in an aqueous environment. In addition, providing a linker-coupled reactant relieves the RNA from the requirement of having substrate properties; the substrate properties are provided by the reactant, while the RNA part of the molecules just has to act as a catalyst. As no nucleotides are attached directly to the reactant, chances are improved that kinetics are not determined by base-pairing interactions.

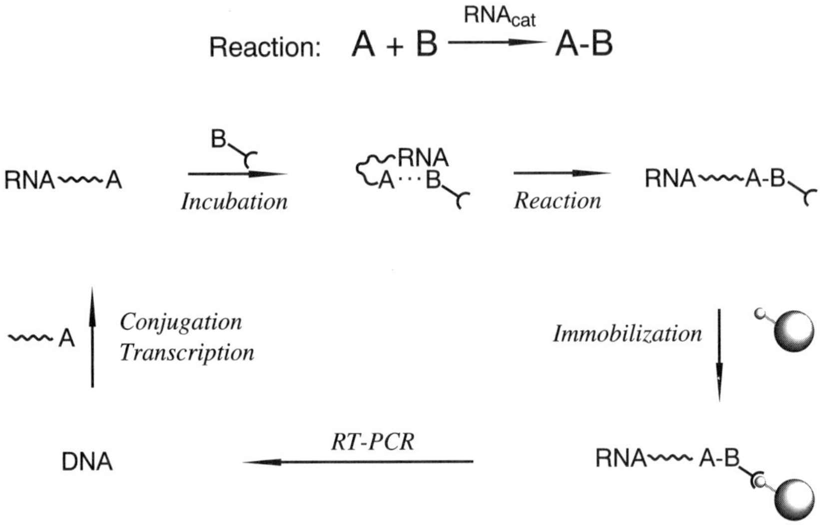

Fig. 2 Direct selection with linker-coupled reactants.

The linker has a number of requirements to meet. After reaction, it should provide a covalent link between the RNA and the anchor group. The linker should be long enough to be unrestrictive in the formation of RNA–substrate complexes and to allow the free orientation of the two reactants with respect to each other and to the RNA. It should be highly flexible and water soluble, ideally it should not provide any features for specific recognition by RNA, and it should show no tendency to the formation of ordered structures. Finally, it should not interfere with the enzymatic amplification reactions. The synthetic polymer polyethylene glycol (PEG) meets these criteria quite well, and its compatibility to oligonucleotide chemistry is well documented (for a recent review, see Jäschke, 1997). Both length and attachment site of the linker are expected to be critical, as with RNA conjugate libraries only part of the 'genetic' information of the conjugate (that is, the nucleotide sequence of the RNA part) can be transferred through the enzymatic amplification steps. Therefore, the non-transferable parts of information, namely the site of attachment and length of the linker, must be kept constant to ensure equivalence of the structures in successive generations. Due to these requirements, the use of the two termini of RNA transcripts seems the most straightforward approach for site-specific derivatization.

The optimum length of the linker is hard to estimate. On the one hand, the extended length of a PEG_{2000} chain (about 45 CH_2CH_2O units) is 160 Å, while the mean end-to-end distance in a coiled conformation is only about 25 Å (Wong et al., 1997). A linker which is too short may be restrictive for ribozymes, where the catalytic centre is far away from the attachment site of the linker, while a linker which is too long might lead to an increased number of intermolecular reactions, causing selection of the wrong species. On the other hand, the successful selection of numerous self-modifying RNAs without using any tethers at all (e.g. Wilson and Szostak, 1995) suggests that the length of the linker might be less important than the fact that there is some spacer that prevents the RNA from recognizing nucleotide features in the direct vicinity of the reactant. We are currently experimenting with linkers composed of 12 to 40 ethylene glycol units.

Methods that so far we have found suitable for the site-specific incorporation of linker-coupled reactants include transcription initiation for 5'-attachment, and enzymatic ligations with both T4 DNA ligase and T4 RNA ligase for specific 3'-coupling. The initiator nucleotide **1**, synthesized for the selection of RNA molecules that catalyse a Diels-Alder reaction between (RNA-coupled) anthracene and (biotinylated) maleimide according to the general scheme (Fig. 2) was found to be incorporated with about 80% efficiency into RNA transcripts, and the conjugates exhibit a low but measurable background reaction with biotin maleimide. Immobilized RNA–linker–product–biotin conjugates are efficiently reverse transcribed and then PCR-amplified (Seelig and Jäschke, submitted).

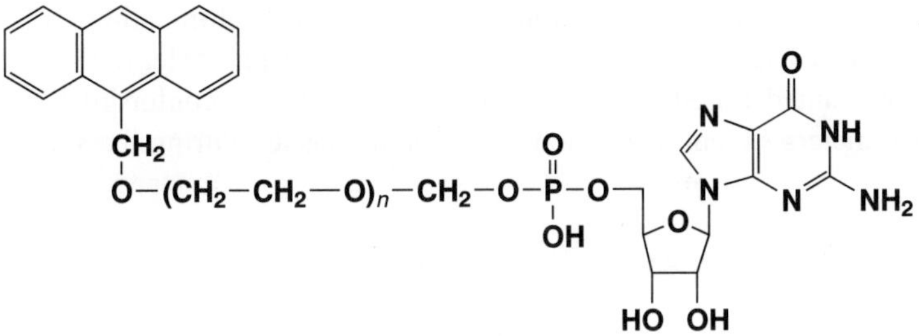

1 $n = 5...40$

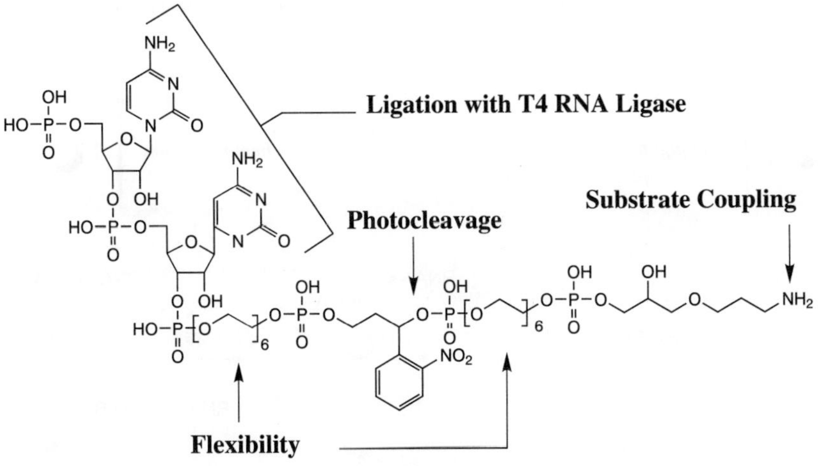

Ligation with T4 RNA Ligase

Photocleavage

Substrate Coupling

Flexibility

2

Besides the obvious advantages of expanding the chemistry and relieving the RNA from the requirement of having substrate properties, the use of a long flexible linker offers an additional advantage that does not exist in traditional direct selection. As already pointed out, the selection criterion is just the covalent attachment of the anchor group to the RNA, no matter where this occurs. As long as the products can be reverse transcribed, they are carried through the SELEX process. There are numerous examples where all or part of the selected species attached the substrate at a position different from the one intended, or even carried out a different chemistry (Illangasekare *et al.*, 1995; Wilson and Szostak, 1995). This phenomenon is an inherent problem of direct selection and will certainly occur in a similar way in selection with linker-coupled reactants. There will be some species where reactant B (Fig. 2) reacts with internal positions of the RNA rather than with the linker-coupled reactant A, and these species may also be carried through the selection process. However, these undesired by-products could be eliminated if a cleavage site is incorporated into the linker (Fig. 3). After immobilization of the reaction products, the RNA species fall into two categories: those that are attached directly to the beads, and those attached via the linker. After cleavage of the linker, only the second group will be released, eliminating catalysts for modification reactions of RNA. We have found that photocleavable linkers are particularly suited for SELEX experiments because they confer high stringency as all parameters except irradiation can remain constant during the selection.

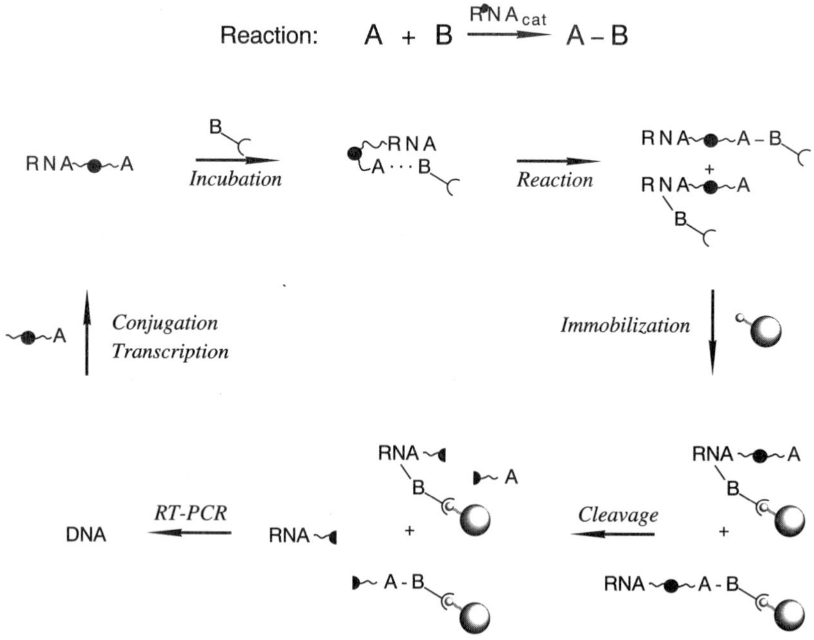

Fig. 3 Direct selection using reactants attached via a cleavable linker.

The dinucleotide analogue **2** was designed containing the following elements:

(i)   a 5′-phosphorylated dinucleotide (pCpC) for enzymatic ligation with T4 RNA ligase (Igloi, 1996);
(ii) two to five hexaethylene glycol units as flexible linkers;
(iii) an O-nitrobenzyl derivative as a photolytic cleavage site (Ordoukhanian and Taylor, 1996); and
(iv) a primary aliphatic amino group for coupling of potential reactants by activated ester chemistry.

The use of this analogue, which was synthesized by phosphoramidite chemistry, proved highly efficient for the incorporation of a variety of reactants into RNA transcripts, and RNA molecules photoreleased after immobilization were found to be intact and amplifiable by RT-PCR (Hausch and Jäschke, submitted).

While the strategies described above (Figs 2 and 3) have focused on the catalysis of reactions involving the formation of covalent bonds, similar schemes are supposed to work for bond cleavage too, and the loss of a characteristic anchor group could be utilized in the partitioning step (Wecker *et al.*, 1996). More complicated to detect are changes that are not associated with the formation or cleavage of covalent bonds, like typical redox reactions. As no anchor group can be introduced during the reaction, one has to use the different reactivity of the linker-coupled reduced form, compared to the oxidized form.

4 Future directions

In the previous paragraph, SELEX with linker-coupled reactants was described using anchor groups introduced or removed during the self-modification step. While this approach allows detection of bond formation and bond cleavage, the application to other types of reaction is problematic. One solution to this problem could be the use of immobilized antibodies that can distinguish between educts and products. Such a strategy would have the additional advantage of providing a much higher stereo- and regioselectivity for distinguishing between the desired product and the products of side reactions. Thus, it is conceivable to raise and immobilize an antibody against one product enantiomer, prepare an RNA pool with linker-coupled reactant, that – in the course of a self-modifying reaction – would produce a racemic product mixture, and then use the immobilized antibody to selectively enrich those RNAs that are coupled to the one enantiomer. Theoretically, even the use of an immobilized aptamer with high specificity for distinguishing between different products would be possible; however, a number of practical issues, like enrichment of RNA sequences on the basis of partial sequence complementarity, would have to be solved.

So far, direct selection for a certain reaction mechanism from random pools has been impossible, since the selection criterion is just a covalent linkage between RNA and anchor group (or the altered electrophoretic or chromatographic mobility). A recent study used well-known mechanistic inhibitors of protein enzymes for the selection of catalytic antibodies that follow a defined mechanism from combinatorial antibody libraries (Janda *et al.*, 1997). A similar scheme is conceivable using RNA libraries. Literally thousands of mechanism-based inhibitors of enzymatic reactions are known, and using these could probably provide an alternative selection strategy that focuses more attention on how catalysis is accomplished. Another way of trapping a reactive intermediate would be the capture of radicals in reactions by using suitably derivatized radical scavengers. Both of these strategies, however, will have to cope with the challenge that the highly reactive intermediates might lead to extensive modifications of the RNA part that are not tolerated by the reverse transcriptase in the amplification step.

The use of coenzymes by RNA will further expand the chemistry of RNA catalysis. A number of aptamers have been described that specifically bind to non-metal cofactors, but so far no example exists where these cofactors were actually used to perform the chemistry of catalysis (Burgstaller and Famulok, 1994; Lauhon and Szostak, 1995). The existence of nucleotide parts in most coenzymes which are irrelevant to the chemistry has repeatedly been hypothesized to stem from an RNA world where the nucleotide part was used as a recognition handle. The application of linker-coupled non-nucleotide reactants might increase prospects for selection of catalysts for reactions involving coenzymes, too.

Current experiments in a number of laboratories are aimed at exploring the catalytic potential of RNA, and the results are often being related to the question of whether RNA is able to perform all the functions of a primitive living system. It is generally accepted that a key enzyme in a hypothetical RNA world is an RNA molecule that can replicate and amplify RNA molecules, and recent results demonstrate that RNA is able to catalyse the polymerization of nucleoside triphosphates on matrix strands (Ekland and Bartel, 1996). Currently, there is no conclusive concept that would explain how such a molecule could have evolved under the anticipated conditions (Joyce and Orgel, 1993). Once the system arrived at this stage, the evolution of other catalytic activities is much easier to imagine. The results from *in vitro* evolution experiments show that RNA molecules can perform much broader catalytic functions than just transesterification of inter-nucleotide phosphodiester bonds. We know that RNA can catalyse the formation of various different types of bonds, but due to the lack of suitable selection methods most of these reactions have been found so far only for self-modification reactions of RNA. The major focus of our work is the development and application of methods to identify catalysts for reactions between two small substrates. With a motivation to investigate prebiotic circumstances, several questions are currently being explored: can RNA catalyse the synthesis of nucleotides from nucleobases and activated sugar phosphates? is RNA able to form heterocyclic ring systems? can RNA form C–C bonds, and can RNA synthesize biologically active cofactors?

As in the case of catalytic antibodies, it is questionable whether catalytic RNA molecules will find practical use for synthetic purposes in the near future. It is, of course, fascinating to create catalysts with designed substrate specificity. Especially in multistep organic synthesis, it would be highly desirable to have catalysts that work with the kind of synthetic targets encountered there. The excellent discrimination between enantiomers or chemically very similar targets (Geiger *et al.*, 1996; Jenison *et al.*, 1994; Klußmann *et al.*, 1996) observed with ground-state molecules will very likely also occur with transition states, and the use of highly specific catalysts would help to avoid complex protection/deprotection or masking schemes and thereby shorten synthetic strategies. However, the high cost of manufacturing and low turnover numbers, long development times, poor stability and the fact that most organic syntheses are carried out in solvents that are incompatible with the use of RNA are only a few of the factors that have to be taken into account, even if RNA would exhibit a catalytic bandwidth sufficient for these kinds of applications.

Acknowledgement

Work in the author's laboratory is supported by the Deutsche Forschungsgemeinschaft.

References

Bartel, D. P. and Szostak, J. W. (1993). *Science* **261**, 1411–1418.
Burgstaller, P. and Famulok, M. (1994). *Angew. Chem. Int. Ed. Engl.* **33**, 1084–1087.
Conn, M. M., Prudent, J. R. and Schultz, P. G. (1996). *J. Am. Chem. Soc.* **118**, 7012–7013.
Ekland, E. H. and Bartel, D. P. (1996). *Nature* **382**, 373–376.
Ellington, A. D. and Szostak, J. W. (1990). *Nature* **346**, 818–822.
Geiger, A., Burgstaller, P., von der Eltz, H., Roeder, A. and Famulok, M. (1996). *Nucleic Acids Res.* **24**, 1029–1036.
Gold, L., Polisky, B., Uhlenbeck, O. and Yarus, M. (1995). *Annu. Rev. Biochem.* **64**, 763–797.
Haldane, J. B. S. (1930). In *Enzymes*, pp. 182. Longmans, Greene, London.
Igloi, G. (1996). *Anal. Biochem.* **233**, 124.
Illangasekare, M., Sanchez, G., Nickles, T. and Yarus, M. (1995). *Science* **267**, 643–647.
Janda, K. D., Lo, L.-C., Lo, C.-H. L., Sim, M.-M., Wang, R., Wong, C.-H. and Lerner, R. A. (1997). *Science* **275**, 945–948.
Jäschke, A. (1997). In *Chemistry and Biological Applications of PEG* (J. M. Harris and S. Zalipsky, eds), ACS Books, in press.
Jenison, R. D., Gill, S. C., Pardi, A. and Polisky, B. (1994). *Science* **263**, 1425–1429.
Jensen, K. B., Atkinson, B. L., Willis, M. C., Koch, T. H. and Gold, L. (1995). *Proc. Natl. Acad. Sci. USA* **92**, 12220–12224.
Joyce, G. F. and Orgel, L. E. (1993). In *The RNA World* (R. F. Gesteland and J. F. Atkins, eds), pp. 1–25. Cold Spring Harbor Laboratory Press.
Klußmann, S., Nolte, A., Bald, R., Eardmann, V. A. and Fürste, J. P. (1996). *Nature Biotechnology* **14**, 1112–1115.
Lauhon, C. T. and Szostak, J. W. (1995). *J. Am. Chem. Soc.* **117**, 1246–1257.
Lerner, R. A., Benkovic, S. J. and Schultz, P. G. (1991). *Science* **252**, 659–667.
Li, Y. and Sen, D. (1996). *Nature Struct. Biol.* **3**, 743–747.
Liu, F. and Altman, S. (1994). *Cell* **77**, 1093–1100.
Lorsch, J. R. and Szostak, J. W. (1994). *Nature* **371**, 31–36.

Lorsch, J. R. and Szostak, J. W. (1996). *Acc. Chem. Res.* **29**, 103-110.

Morris, K. N., Tarasow, T. M., Julin, C. M., Simons, S. L., Hilvert, D. and Gold, L. (1994). *Proc. Natl. Acad. Sci. USA* **91**, 13028-13032.

Ordoukhanian, P. and Taylor, J. S. (1996). *J. Am. Chem. Soc.* **117**, 9570-9571.

Pauling, L. (1946). *Chem. Eng. News* **24**, 1375.

Prudent, J. R. and Schultz, P. G. (1996). In *Catalytic RNA* (F. Eckstein and D. M. J. Lilley, eds), pp. 383-395. Springer, Berlin.

Prudent, J. R., Uno, T. and Schultz, P. G. (1994). *Science* **264**, 1924-1927.

Smith, D., Kirschenheuter, G. P., Charlton, J., Guidot, D. M. and Repine, J. E. (1995). *Chem. Biol.* **2**, 741-750.

Tuerk, C. and Gold, L. (1990). *Science* **249**, 505-510.

Wecker, M., Smith, D. and Gold, L. (1996). *RNA* **2**, 982-994.

Williams, K. P. and Bartel, D. P. (1996). In *Catalytic RNA* (F. Eckstein and D. M. J. Lilley, eds), pp. 367-381. Springer, Berlin,

Wilson, C. and Szostak, J. W. (1995). *Nature* **374**, 777-782.

Wong, J. C., Kuhl, T. C., Israelachvili, J. N., Mullah, N. and Zalipsky, S. (1997). *Science* **275**, 820-822.

Conclusions and Opportunities

BRIAN W. METCALF
SENIOR VICE PRESIDENT,
CHEMICAL AND CELLULAR SCIENCES
SmithKline Beecham Pharmaceuticals, 709 Swedeland Road, King of Prussia, PA 19406, USA

The systematic exploitation of RNA as a drug target is a new field and as such is relatively immature. This immaturity was reflected during the symposium by the paucity of experiments at the level of whole cells or in the description of *in vivo* results. Practitioners in the field were urged to extend their experimentation from *in vitro* observations to that in whole cells in order to underline the importance of RNA as a drug target. In so doing, judicious choice of targets should be undertaken to select those with a high degree of likelihood of demonstrating biomedical utility.

The exploitation of secondary RNA motifs offers an under-explored opportunity to identify novel drug targets. Such exploitation is attractive owing to the many faces of RNA:

- RNA demonstrates a propensity to undergo induced secondary conformations on binding of ligands. A ligand that interacts with a specific RNA secondary motif might induce the target RNA to undergo a conformational change so that it is no longer able to perform its natural function. Ligands that interact in this manner would be uncompetitive inhibitors versus the endogenous ligand and would not have to directly displace the endogenous ligand.

- RNA has many functions ranging from encoding protein sequence to structural roles to catalytic function to regulatory control. Sometimes the same RNA sequence will have a dual function such as structural support and an encoding purpose.

- In an infectious disease setting, a target with dual functions may undergo mutations to generate resistance to a targeted ligand at a much slower rate than single function targets.

THE MANY FACES OF RNA
ISBN 0-12-233210-5

Impact on biomedical research

This observer suggested several areas that he feels offer clear pathways towards demonstrating the biomedical utility of RNA-based approaches including:

- the use of 'designer' ribozymes for the functional characterization of novel genes in cell culture and *in vivo*. The explosion of available gene sequences arising from high throughput sequencing now demands subsequent methodology to select the most appropriate genes and gene products as targets for drug discovery.
- Exploitation of the observation that the antitumour agent bleomycin induces RNA cleavage, as well as DNA cleavage. Perhaps the antitumour activity arises from RNA attack.
- Exploitation of RNA motifs in infectious disease settings. RNA viruses and retroviruses appear to offer many opportunities. Bacteria use an essential ribozyme, RNAse P, to process precursor t-RNA. In addition, the molecular target for the aminoglycoside antibiotics such as paromomycin has been shown to be a secondary RNA motif in susceptible bacteria. As described at this symposium, the binding site for an aminoglycoside has been elucidated and binding interactions defined at the atomic level by NMR spectroscopy (Fourmy *et al.*, 1996). Interactions which have been identified between paromomycin and RNA are reminiscent of ligand binding to protein sites and suggest that rational design of ligands that bind to RNA will be possible.

The nature of RNA, and hence of ligands such as the aminoglycosides that bind to it, may be self limiting from a small molecule drug perspective. That is, one may assume that ligands for RNA may need to carry several positive charges in order to complement the negatively charged RNA. Such charged species may well have liabilities in cell penetration and in pharmacokinetics that would manifest themselves as short half-lives *in vivo* and in lack of oral activity. On the other hand, similar issues have been addressed in other fields such as integrin antagonism where the natural ligand (RGD) presents an arginine residue. Solutions to the pharmacokinetic issues in this area have been devised and 'arginine mimetics' may find use as RNA-directed ligands.

It is felt that retroviruses such as HIV and RNA viruses such as hepatitis C offer the most tractable and accessible targets for an RNA-directed approach. In contrast to antibacterials and antifungal agents, RNA ligands directed at viral RNA will have to penetrate human cells rather than the less penetrable bacterial or fungal cells. The RNA genome of HIV would appear to offer many likely targets, and, therefore, it is not surprising that there have been some reports of successful interruption of viral function by small molecules directed at secondary RNA motifs. For example, Zapp *et al.* (1993) describe the interaction of an aminoglycoside antibiotic, neomycin, with the HIV Rev Responsive Element (RRE). Neomycin thereby blunts the production of HIV from chronically infected cells. Another example, reported at the time of this symposium in abstract form, but since in the refereed literature, is the design of small molecules which block the binding of the HIV transactivating protein, Tat, to its responsive RNA element TAR (Klimkait *et al*., 1996; Hamy *et al*., 1997). This competition with Tat results in an anti-HIV effect in cell culture.

The administration of α interferon is the only available treatment for the RNA virus hepatitis C (HCV), but poses a number of disadvantages including route of administration, side effect profile and limited efficacy. Hence, HCV offers the potential of a number of RNA targets as the 3' and 5' NTRs that contain regulatory signals. Presumably, such NTRs interact with viral or cellular proteins. Delineation of the minimal interacting sequences would provide a basis for the design of molecular screens to find small molecules that interfere with these interactions and hence with regulation of HCV expression.

The DNA viruses, such as those of the Herpes family, also offer opportunities to exploit RNA. For example, during its latency periods, HSV-2 expresses RNA sequences known as latency-associated transcripts (LATs). While the role of these LATs is not understood and may be related to an antisense mechanism of translational arrest, interactions with viral and cellular proteins may also be involved. It would appear that interference of the binding of such LATs with endogenous cell proteins, if identified, may prevent establishment of latency, or block viral reactivation from latency.

Progress in the exploitation of RNA in biomedical applications will depend critically on the choice of RNA target. These targets should be chosen so that interacting ligands have the best chance of success in that inevitable problems of pharmacokinetics, cell penetration or cellular efflux, and selectivity will be minimized by the choice. There appear to be many opportunities. It is incumbent on those working in the field and on those about to exploit the many faces of RNA to take the currently existing series of tantalizing observations and extend them to the next step in demonstrating the biomedical relevance of interruption of RNA function.

References

Fourmy, D., Recht, M. I., Blanchard, S. C. and Puglisi, J. D. (1996). *Science* **274**, 1367–1371.

Hamy, F., Felder, E. R., Heisman, G., Lazdins, J., Aboul-Ela, F., Varani, G., Karn, J. and Klimkait, T. (1997). *Proc. Natl. Acad. Sci. USA* **94**, 3548–3553.

Klimkait, T., Heizmann, G., Felder, E., Varani, G., Karn, J., Lazdins, J. and Hamy, F. (1996). *Antiviral Research* **30**, A17.

Zapp, M. L., Stern, S. and Green, M. R. (1993). *Cell* **74**, 969–978.

Appendix: Abstracts of Posters

Contents

Synthesis and Properties of RNA-Oligonucleotides Containing 6'-O-functionalized Allofuranosyl Nucleosides. Xiaolin Wu and Stefan Pitsch. *Organisch-chemisches Laboratorium der Eidgenössischen Technischen Hochschule, Universitätstr. 16, CH-8092 Zürich*

A general and short synthesis of the phosphoramidite building block **II** from D-glucose and its efficient incorporation into RNA-oligonucleotides **I** (using solid phase methodology on a DNA-synthesizer) is presented, together with a discussion of the pairing properties of such functionalized nucleic acids.

Selective Pd^0-assisted removal of the allyloxycarbonyl group [1] from the fully protected oligonucleotide sets the aminopropyl group free, which will be further modified, still taking advantage of solid phase chemistry.

In the course of this project, a new method for the selective introduction and the efficient removal of the photocleavable *ortho*-nitrobenzyloxymethyl group (*NBM*) [2] was developed. This group has been chosen for the protection of all (natural and modified) nucleosides.

(I)   (II)

[1] Hayakawa, Y., Hirose, M. and Noyori, R. (1993). *J. Org. Chem.* **58**, 5551.
[2] Schwartz, M. E., Breaker, R. R., Asteriadis, G. T., deBear, J. S. and Gough, G. R. (1992). *BioMed. Chem. Lett.* **9**, 1019.

Incorporation of Modified Nucleotides into RNA: Useful Probes for Structure–Function Studies.

James B. Murray, Chris J. Adams, Tim Moss, Peter G. Stockley and John R. P. Arnold. *Department of Biology, University of Leeds, Leeds LS2 9JT, UK*

The phosphoramidite derivatives of 4-thiouridine, 6-thioguanosine, 6-thioinosine and 2-pyrimidinone ribonucleoside (a fluorescent nucleoside) have been synthesized and incorporated into oligoribonucleotides up to 50 nucleotides long. The thio-containing bases are photoreactive and have been used in photo-cross-linking studies. In addition, these bases introduce subtle modifications for probing the roles of exocyclic carbonyl and amino groups.

These properties have been used to examine the roles of conserved nucleotides in the central core of the hammerhead ribozyme. The kinetic properties of the variants described here are consistent with several key interactions seen in hammerhead crystal structures, in particular they provide experimental support for the assignment of the proposed catalytically active magnesium-binding site.

Modified Nucleotides in NMR Studies of RNA Structure and Dynamics.

John R. P. Arnold[1], Julie Fisher[2], James B. Murray[1] and A. Kay Collier[2]. [1]*Department of Biology;* [2]*School of Chemistry, University of Leeds, Leeds, LS2 9JT, UK*

RNA structure can be investigated using ^{1}H NMR as can be seen from the number of motifs (relatively small stem-loops, helices, etc.) which have now been characterized in this way, often yielding coarse to medium resolution structural information. But with increasing molecular weight, crowding in the ^{1}H spectrum soon becomes prohibitive. Uniform or nucleotide-specific ^{13}C/^{15}N labelling extends the range of RNA systems that NMR techniques can be applied to. A large number of possible coherence transfer pathways become available, and there are now many multi-dimensional multinuclear experiments to exploit this. However, the procedures for preparing labelled RNA are laborious and expensive, even when labelled NTPs are commercially available.

An alternative approach is the incorporation of chemical modifications, which can be done site-specifically with chemical synthesis. Modifications can introduce unique NMR signatures to one or more specific sites in an RNA molecule. Obviously the information that can be conveyed from a small number of such signals may in some respects be limited, but specific questions concerning RNA structure and dynamics can be readily addressed. The resulting spectra are generally easy to analyse.

A potential complication is that any modification to the RNA may alter its structure and function, complicating any conclusions with respect to unmodified RNA. However, often an assay of one sort or another is available to help resolve this.

Here we give some examples where chemical modifications have been applied to questions of both RNA structure and dynamics. These have provided data which would have been much more difficult to obtain using other NMR strategies.

[1]   Murray, J. B., Collier, A. K. and Arnold, J. R. P. (1994). *Analytical Biochem.* **218**, 177–184.
[2]   Collier, A. K., Arnold, J. R. P. and Fisher, J. (1996). *Magnetic Resonance in Chemistry* **34**, 191–196.

Oligonucleotide Purification. Helen Birrell[1], Victoria Emerick[2] and Neville Nicholson[1]. [1]*SmithKline Beecham Pharmaceuticals, Harlow, UK;* [2]*SmithKline Beecham Pharmaceuticals, Upper Providence, USA*

Polyacrylamide slab gel electrophoresis is widely used for the purification of milligram quantities of oligonucleotides. The technique gives good resolution, but the required band must be cut from the gel and eluted from the gel matrix which is time consuming and results in contamination of the product with polyacrylamide decomposition products. The process is capable of purifying only small batches of the required oligonucleotide and gives poor recoveries.

To ameliorate these disadvantages, we have used a gel in the form of a column rather than a slab, from which the electrophoresis products are continuously eluted and collected in a flow of buffer. The apparatus was designed by Bio-Rad for the purification of proteins and termed Prep Cell 491. We have used it for the purification of a chemically synthesized 39mer DNA and monitored the separation by gel-filled capillary electrophoresis. Some fractionation occurred, but was insufficient to provide complete removal of close running impurities. A recent publication (*Nucleic Acids Research* (1996), **24**(18), 3647) describes similar results for RNA and we and the authors of the publication are co-operating with Bio-Rad to extend these findings.

RNA Sequencing by Digestion with Sequence-specific Ribonucleases followed by MALDI-ToF. David A. Tolson and Neville Nicholson. *Computational and Structural Sciences, SmithKline Beecham Pharmaceuticals, New Frontiers Science Park, Harlow CM19 5AW, Essex, UK*

Classical methods of RNA sequencing using either chemical or enzymatic degradation are labour intensive, time consuming, and involve the use of radioisotopes. Enzymatic digestion followed by mass spectrometry offers the potential of a simple, sensitive and rapid method of oligonucleotide sequencing. DNA sequencing using snake venom phosphodiesterase (SVP) and bovine spleen phosphodiesterase (BSP), which sequentially digest the DNA from 3′ end or the 5′ end respectively, followed by matrix-assisted laser desorption ionization time-of-flight mass spectrometry (MALDI-ToF) has been reported.

Unambiguous assignment of the DNA bases is readily achieved due to the minimum 9 Da mass difference between the nucleotides dA and dT, which is easily resolved by the MALDI-ToF instrument. When the same procedure is applied to RNA, it is not possible to unambiguously assign the nucleotides U and C due to their 1 Da mass difference. Here we report our experiences with sequence-specific ribonuclease digestion followed by MALDI-ToF in order to distinguish between the pyrimidine bases of RNA.

Computer Simulation of RNA Folding by a Genetic Algorithm.

Alexander P. Gultyaev[1,2], F. H. D. van Batenburg[2] and Cornelis W. A. Pleij[1]. [1]*Leiden Institute of Chemistry, Leiden University, PO Box 9502, 2300 RA Leiden, The Netherlands;* [2]*Section Theoretical Biology, EEW of Leiden University, Kaiserstraat 63, 2311 GP Leiden, The Netherlands*

Functioning of RNA molecules strongly depends on their structures. The folding of RNA structure is a dynamic process that occurs in a complicated landscape determined by large variety of different shapes. Therefore the final structure and the time of its folding may depend on the specific folding pathway.

A procedure for simulating the RNA folding process using the principles of genetic algorithms has been proposed [1,2]. The algorithm simulates a folding pathway of RNA as a stepwise process with formation and disruption of temporarily formed intermediate structures. This provides an opportunity to implement kinetic features of RNA folding into simulation, where energetic barriers of elementary steps are determined by free energies of loops and helical stacks in folding intermediates. An important kinetic effect of RNA folding during its synthesis is included in the algorithm as well. In contrast to modern algorithms of predicting the secondary structure with the lowest energy, the stepwise folding simulation allows to predict RNA pseudoknot formation [3].

It is shown that the folding pathway simulation can result in structure predictions that are more consistent with proven structures than minimal energy solutions. This demonstrates that RNA folding kinetics is very important for the formation of functional RNA structures. In addition, the simulations are able to predict functional metastable foldings and kinetically driven transitions to more stable structures. For example, a metastable structure of primer RNA is shown to be involved in the copy number regulation of the ColE1 group plasmids [4]. Not only RNA structures are important for the function, but the pathways and the timescale of refolding processes are essential, and examples of simulations for different types of RNAs (e.g. plasmid-coded regulatory RNAs, viroid RNA) show that the proposed approach can be a powerful tool in the studies of RNA folding.

[1] van Batenburg, F. H. D., Gultyaev, A. P. and Pleij, C. W. A. (1995). *J. Theor. Biol.* **174**, 269–280.

[2] Gultyaev, A. P., van Batenburg, F. H. D. and Pleij, C. W. A. (1995). *J. Mol. Biol.* **250**, 37–51.

[3] Gultyaev, A. P., van Batenburg, F. H. D. and Pleij, C. W. A. (1994). *J. Gen. Virol.*
     75, 2851–2856.
[4] Gultyaev, A. P., van Batenburg, F. H. D. and Pleij, C. W. A. (1995). *Nucleic Acids
     Res.* **23**, 3718–3725.

Solution Studies of the Dimerization Initiation Site (DIS) of HIV-1 Genomic RNA. Frédéric Dardel[1], Manuela Cuffiani[1], Chantal Ehresmann[2], Bernard Ehresmann[2] and Sylvain Blanquet[1]. *[1]Laboratoire de Biochimie, URA 1970 du CNRS, Ecole Polytechnique, 91128 Palaiseau Cedex, France; [2]UPR 9002 du CNRS, IBMC, 15 rue René Descartes, 67084, Strasbourg cedex, France*

HIV-1 has a diploid genome composed of two identical RNA molecules. These two RNA strands are bound through a small region close to their 5' terminus. This dimerization is essential to several steps of the viral replicative cycle. Initiation of dimerization involves a conserved stem-and-loop structure around position 275 (DIS). This stem-and-loop structure features a palindromic sequence within the loop and has been shown to mediate dimerisation via a 'kissing-complex' involving loop–loop interactions.

We have investigated the capacity of small oligoribonucleotides to dimerize *in vitro*, in order to define a minimal DIS sequence, amenable to NMR structural determination. A 19-mer RNA sequence (DIS-19) has been defined which forms a stable symmetric dimer in solution at millimolar concentrations, even in the absence of Mg^{2+}. RNA ligation experiments strongly suggest that DIS-19 forms a stable kissing complex under the NMR experimental conditions and preliminary NMR data indicate that the helix formed by loop–loop base pairing is coaxial with the stem. A uniformly ^{15}N labelled sample of DIS-19 has been produced and experiments using $^{14}N/^{15}N$ heterodimers are currently undertaken in order to demonstrate the loop–loop interaction.

Supported by the Agence Nationale de Recherche sur le Sida.

Cross-linking Studies of the HIV-1 Regulatory Protein Tat to Synthetic TAR RNA. Mark A. Farrow, Nikolai N. Naryshkin and Michael J. Gait. *Medical Research Council, Laboratory of Molecular Biology, Hills Road, Cambridge, CB2 2QH, UK*

We have been studying the interactions of HIV-1 Tat with its target RNA the *trans*-activation response element (TAR) by the introduction of reactive groups at defined positions within a synthetic TAR RNA model duplex [1]. It has previously been shown that a synthetic Tat peptide (ADP-1) binds with high affinity and specificity to the TAR

RNA [2]. Upon formation of a complex between TAR and the Tat peptide, the reactive groups within the RNA form covalent bonds with the amino acids which are in close proximity to them. It is possible to purify the covalent RNA–peptide complexes away from the non-covalent complexes and any unreacted material by HPLC. The amino acids which are covalently linked to the RNA can then be identified by proteolysis using a number of different enzymes followed by mass spectroscopy. Our aim is to define which amino acids are in close proximity to particular positions within the RNA, to orientate the peptide within the RNA–peptide complex, and to assist in the development of a 3-D model of the complex in conjunction with parallel NMR studies of the Tat TAR complex [3] (in collaboration with F. Aboul-ela and G. Varani, LMB Cambridge).

[1]   Hamy, F., Asseline, U., Grasby, J., Shigenori, I., Pritchard, C., Slim, G., Butler, P. J., Karn, J. and Gait, M. J. (1993). *J. Mol. Biol.* **230**, 111–123.
[2]   Churcher, M. J., Lamont, C., Hamy, F., Dingwall, C., Green, S. M., Lowe, A. D., Butler, P. J., Gait, M. J. and Karn, J. (1993). *J. Mol. Biol.* **230**, 90–110.
[3]   Aboul-ela, F., Karn, J. and Varani, G. (1995). *J. Mol. Biol.* **253**, 313–332.

Translational Control by Kinetics of RNA Folding. J. van Duin, H. Groeneveld, K. Thimon, R. Poot and I. Boni. *Leiden Institute of Chemistry, Department of Biochemistry, Leiden University, The Netherlands*

The gene for the maturation (A) protein of the single-stranded RNA coliphage MS2 is preceded by an untranslated leader of 130 nt which folds into a cloverleaf, i.e. three stem-loop structures enclosed by a long-distance interaction (LDI). This LDI is essential for translational control. Its 3′ moiety contains the Shine-Dalgarno region, whereas its complement is located 80 nt upstream. Mutational analysis shows that this base pairing represses expression of the A-protein gene. We present a model in which translational starts can only take place on nascent, non-equilibrated RNA, in which base pairing between the complementary regions has not yet taken place. We suggest that this pairing is kinetically delayed by the intervening sequence, which contains the three hairpins of the cloverleaf. The model is based on several observations. Removing the intervening sequence reduces expression, whereas increasing its length has the opposite effect. In addition, further stabilization of the LDI by a stronger base pair does not lead to a decrease in A-protein synthesis. Such a decrease is predicted to occur if translation would be controlled by the equilibrium structure of the leader RNA [1]. *In vitro* experiments show that ribosome binding to the start of the A-protein gene is faster than the renaturation of denatured wild-type MS2 RNA. However, MS2 RNA from which the intervening sequence has been deleted renatures faster than ribosomes can bind. These and other observations fit a kinetic model of translational control by RNA folding.

Delays in folding of RNA are believed to be caused by temporal trapping of the chain in metastable intermediates. We are presently identifying these intermediates, which form the kinetic barrier to the equilibrium folding.

[1]   Groeneveld, H., Thimon, K. and van Duin, J. (1995). *RNA* **1**, 79–88.

Modelling of the Three-dimensional Architecture of Bacterial Ribonuclease P RNAs Based on Comparative Sequence Analysis.

Christian Massire, Luc Jaeger, Eric Westhof. *Institut de Biologie Moléculaire et Cellulaire du CNRS, 15 rue Descartes, 67084 Strasbourg Cedex, France*

The secondary structure of bacterial RNase P RNA sequences can be classified into two types: type A (e.g. *Escherichia coli* [1]) or type B (e.g. *Bacillus subtilis* [2]). Both share a common catalytic core formed by the assembly of two self-folded domains but differ in their peripheral elements. A new alignment of 112 sequences from the RNase P database [3] reveals several instances of covariations, that allow the refinement of the known secondary structure and the finding of new tertiary contacts. Some of the new interactions are involved in the association of the two core domains, whereas some are more specific to the peripheral extensions typical of either type. With the help of molecular modelling, a three-dimensional model was built for each type. In each model, the core domains create a cleft in which the ptRNA substrate is bound, with most evolutionarily conserved residues converging towards the cleavage site. The inner cores of both types are stabilized similarly albeit by different peripheral elements, emphasizing the modular and hierarchical organization of the architecture of RNase P RNAs. Other noteworthy features of these 3D models include their compactness and their good agreement with crosslinking experiments. Another family of catalytic RNA molecules, the group I introns, exploits also different structural elements located in similar spatial positions in related molecules in order to promote long-range tertiary contacts that are different but which possess identical functions [4].

[1] Brown, J. W., Nolan, J. M., Haas, E. S., Rubio, M. A. T., Major, F. and Pace, N. R. (1996). *Proc. Natl. Acad. Sci.* **93**, 3001–3006.
[2] Haas, E. S., Banta, A. B., Harris, J. K., Pace, N. R. and Brown, J. W. (1996). *Nucleic Acids Res.* **24**, 4775–4782.
[3] Brown, J. W. (1997). *Nucleic Acids Res.* **25**, 263–264.
[4] Lehnert, V., Jaeger, L., Michel, F. and Westhof, E. (1996). *Chem. & Biol.* **3**, 993–1009.

Cloning, Expression and Characterization of *Staphylococcus aureus* Ribonuclease P.

Sabine Guth, Lisa Hegg, and Cathy Prescott. *SmithKline Beecham Pharmaceuticals, RNA Research Group, 1250 S. Collegeville Road, Collegeville, PA 19426-0989, USA*

Ribonuclease P is an essential and ubiquitous ribonucleoprotein enzyme which is involved in the maturation of precursor tRNAs and precursor 4.5 S RNA. To date, every cell and eukaryotic subcellular compartment in which tRNAs are synthesized contain RNase P or demonstrate RNase P activity. *In vitro*, the prokaryotic enzyme demonstrates enzymatic activity in the absence of its protein cofactor. In contrast, the protein component is absolutely required by the human homologue (for review see [1] and [2] and references contained within). We sought to clone, express and characterize RNase P from the clinically relevant bacterium *Staphylococcus aureus*. Random sequencing of a *S. aureus* genomic library followed by sequence homology searching with the Gram positive *Bacillus subtilis* RNase P RNA and protein sequence revealed sequences which encoded both the *S. aureus* RNase P RNA and protein. The full genes were obtained by PCR methods. These sequences were cloned into suitable expression vectors and over-expressed. The optimal conditions for cleavage of pretRNA or a pretRNA analogue were determined for both the *S. aureus* RNase P RNA and the holoenzyme. *In vitro*, the *S. aureus* RNase P RNA was able to cleave RNA substrates under high ionic strength conditions. The holoenzyme can be reconstituted under low ionic strength and low magnesium concentrations. *S. aureus* RNase P represents a target for the development of anti-bacterial agents.

[1]   Kirsebom, L. A. (1995). *Mol. Microbiol.* **17**, 411–420.
[2]   Chamberlain, J. R., Tranguch, A. J., Pagan-Ramos, E. and Engelke, D. R. (1996). *Prog. Nucleic Acids Res. Mol. Biol.* **55**, 87–119.

RNase P: an RNA-based Target for Anti-bacterial Agents.
Michael Gress[1], Neil Pearson[2], Cathy Prescott[1] and Lisa Hegg[1]. *SmithKline Beecham Pharmaceuticals, [1]RNA Research Group, 1250 S. Collegeville Road, Collegeville, PA 19426-0989, USA; [2]Medicinal Chemistry, Harlow, CM19 5AD, UK*

RNA plays an essential role in every cellular macromolecular process. Many natural antibiotics (e.g. spectinomycin) bind and inhibit functional RNA or ribonucleoprotein complexes. Therefore, RNA-based targets represent an exciting area for the discovery of anti-bacterial agents. Ribonuclease P is an essential endoribonuclease required for the maturation of RNA substrates including tRNAs and 4.5 S RNA. The enzyme is a ribonucleoprotein complex containing both RNA and protein components. In prokaryotes this complex has a stoichiometry of 1:1 whereas the eukaryotic homolog contains a higher protein content [1]. Differences between prokaryotic and eukaryotic RNase P make it a specific and novel target for exploitation and anti-bacterial drug development. Much of our current knowledge is based upon studies of RNase P from the Gram negative *Escherichia coli* and Gram positive *Bacillus subtilis* [2,3]. We sought to express the RNase P RNAs from these organisms as well as the RNAse P proteins from *E. coli* and *Staphylococcus aureus* (see S. Guth *et al.*, this meeting) to identify inhibitors of RNase P. Additionally, we investigated puromycin (an analogue of the 3′ end of tRNA) inhibition of RNase P activity. Previous work has shown that *E. coli* RNase P holoenzyme activity is inhibited by puromycin, implying that the CCA of tRNA is important for recognition by RNase P [4]. Interestingly, the *B. subtilis* RNase P appears to be able to recognize substrates regardless of the presence or absence of the CCA 3′ terminus [5]. Evidence of inhibition of RNAse P activity by compounds and puromycin will be presented and implications will be discussed.

[1]   Darr, S. C., Brown, J. W. and Pace, N. R. (1992). *Trends in Biochem. Sci.* **17**, 178–182.
[2]   Altman, S., Kirsebom, L. and Talbot, S. (1993). *FASEB J.* **7**, 7–14.
[3]   Pace, N. R. and Smith, D. (1996). *J. Biol. Chem.* **265**, 3587–3590.
[4]   Vioque, A. (1989). *FEBS Lett.* **246**, 137–139.
[5]   Vold, B. and Green, C. J. (1988). *J. Biol. Chem.* **263**, 14390–14396.

Structural and Thermodynamic Studies of *Bacillus subtilis* RNase P Catalytic Centre.

Julia Hubbard, Lesley Maclachlan, Claus Spitzfaden, Craig Brooks, Neville Nicholson and Drake S. Eggleston. *Department of Computational and Structural Sciences, SmithKline Beecham Pharmaceuticals, New Frontiers Science Park; Harlow CM19 5AW, Essex, UK*

The endoribonuclease RNase P is a ubiquitous and essential ribonucleoprotein. It cleaves the 5′ terminal leader sequences of precursor tRNAs (ptRNAs) to generate mature tRNAs. Much of the current understanding of this enzyme is derived from studies of RNase P from *Escherichia coli* and *Bacillus subtilis*. [1] The solution conformation of the loop structure that constitutes the putative substrate and divalent metal ion(s) binding site in *E. coli* has been studied by NMR [2] to try to gain insight on the function of this ribozyme. In *E. coli*, the substrate binding region is found in an internal loop, whereas in *B. subtilis* the motif is part of a hairpin loop. A 23 mer RNA representing this *B. subtilis* domain has been prepared by chemical synthesis. In order to stabilize the secondary structure with the aim of defining the solution conformation of the functional loop, we have added an additional region of helix at the 5′ and 3′ ends.

Homonuclear ^{1}H NMR spectroscopy has been used to investigate structural and dynamic features of this RNA sequence. Spectra have been acquired in two buffers: 20 mM cacodylic acid, 20 mM NaCl, 0.1 mM EDTA-d12 in D_2O (a buffer reported to preserve the monomeric species in solution) and 100 mM NaCl, 10 mM sodium phosphate, pH 6.5, in H_2O and D_2O. The melting of the RNA sample has been studied in H_2O solution by 1D ^{1}H NMR spectroscopy. 2D homonuclear correlation and nOe spectra have been used to characterize the primary sequence of the RNA and to obtain assignment information. The effect of Mg^{2+} also has been investigated by 1D ^{1}H NMR spectroscopy.

[1]   Hames, B. D. and Glover, D. M., eds (1994). *RNA–protein interactions.* IRL Press, Oxford.
[2]   Glemarec, C., Kufel, J., Foldesi, T., Sandstrom, Kirsebom, L. A. and Chattoppad-hyaya, J. (1996). *Nucleic Acids Res.* **24**, 2022–2035.

Inter-domain Cross-linking and Molecular Modelling of the Hairpin Ribozyme.

David J. Earnshaw[1]*, Benoit Masquida[2], Eric Westhof[2], Snorri Th. Sigurdsson[3], Fritz Eckstein[3], Sabine Müller[1] and Michael J. Gait[1]. [1]*Medical Research Council, Laboratory of Molecular Biology, Hills Road, Cambridge, CB2 2QH, UK;* [2]*IBMC-CNRS, 15 rue René Descartes, F-67084 Strasbourg, France;* [3]*Max-Planck Institüt für Experimentelle Medizin, Hermann-Rein-Strasse 3, D-37075, Göttingen, Germany.* Corresponding author.*

The hairpin ribozyme derived from the negative strand of a satellite RNA of tobacco ringspot virus [(−)s TRSV] can catalyse the efficient cleavage of an external RNA substrate in *trans*. Unlike the hammerhead ribozyme for which crystal structures have already been published, [1,2] the hairpin ribozyme does not have such information regarding its tertiary structure. In order to provide some insight into its 3-D structure, a novel cross-linking technique [3] was used in an attempt to understand how the two domains of the hairpin ribozyme may fold and to construct a preliminary molecular model. This technique involved the site specific incorporation of 2′-amino-2′deoxy pyrimidine nucleosides into a three stranded hairpin ribozyme by RNA automated chemical synthesis. These positions included four within loop A on the substrate strand (U2, C3, U5 and U7) and three positions in the loop B domain on ribozyme strand B (U39, U42, C44). These positions were cross-linked by means of either an aryl or an alkyl disulphide linkage. In total, 22 cross-linked ribozymes were prepared and kinetic parameters were measured under single turnover conditions and compared in parallel with the kinetic parameters for the substrate ribozyme B complex treated with DTT to open the disulphide linkage. The data revealed that cross-links U2/C44, C3/U39, and C3/C44 drastically reduced the catalytic rate, whereas cross-links C3/U42 and U7/C44 resulted in small but significant reductions in catalytic activity. For four of the cross-links U2/U39, U2/U42, U7/U39 and U7/U42 the catalytic rates were unaltered or slightly reduced. These data provided assistance in assembling a molecular model of the 3D-fold of the hairpin ribozyme which incorporates all the previously published chemical and enzymatic probing information.

[1]   Pley, H. W., Flaherty, K. M. and McKay, D. M. (1994). *Nature* **372**, 68–74.
[2]   Scott, W. G., Finch, J. T. and Klug, A. (1995). *Cell* **81**, 991–1002.
[3]   Sigurdsson, S. T., Tuschl, T. and Eckstein, F. (1995). *RNA* **1**, 575–583.

Mechanistic Studies of the Hairpin Ribozyme.

Francesca Gill, Jane A. Grasby and Karen J. Young. *Department of Chemistry, Krebs Institute, University of Sheffield, Sheffield, S3 7HF, UK*

Ribozymes are RNA molecules which have the ability to catalyse the formation and cleavage of phosphodiester bonds. The hairpin ribozyme is a member of a family of small ribozymes which also includes the hammerhead and hepatitis delta. The cleavage reaction catalysed by ribozymes requires the presence of a divalent metal ion (normally

Mg^{2+}) and produces a $2',3'$-cyclic phosphate and a $5'$-hydroxyl terminus. [1,2] It has been suggested that the initial deprotonation of the $2'$-hydroxyl required for the reaction is provided by a solvated divalent metal ion hydroxide. A further role of the metal ion may be that it binds the pentacoordinate $TS^{\ddagger}$. It may also be possible that the metal ion stabilizes the $5'$-oxyanion leaving group, either directly or in a water mediated process. [3]

The role of the metal ion in the reaction has been investigated by the synthesis of substrates containing a phosphorothioate linkage in place of the scissile phosphodiester bond. The values of k_{cat} and K_m for the wild type substrate, and the R_p and S_p phosphorothioate containing substrates, have been determined.

Substrate	K_m (nM)	k_{cat} (min^{-1})	k_{cat}/K_m (M^{-1} min^{-1})	Relative activity*
Wild type	49.4	0.191	3.87×10^6	1.00
R_p epimer	40.0	0.259	6.48×10^6	1.67
S_p epimer	18.4	0.236	1.28×10^7	3.31

$$*\text{Relative activity} = \frac{k_{cat}/K_m \text{ for modified substrate}}{k_{cat}/K_m \text{ for wild type substrate}}$$

The rates of reaction for all three substrates are comparable and they suggest that the metal ion is not involved in an inner sphere mechanism. To further investigate this discovery, the cleavage reaction has also been monitored in the presence of cobalt (III) complexes. The data obtained from these experiments reinforces the notion of an outer sphere mechanism.

[1]   Dahm, S. A. and Uhlenbeck, O. C. (1991). *Biochemistry* **30**, 9464–9469.
[2]   Slim, G. and Gait, M. J. (1991). *Nucleic Acids Res.* **19**, 1183–1188.
[3]   Dahm, S. A., Derrick, W. B. and Uhlenbeck, O. C. (1993) *Biochemistry* **32**, 13040–13045.

Twin-ribozyme-introns: a New Category of Group I Introns with Conserved Structural Organization and Function.

Christer Einvik, Peik Haugen and Steinar Johansen. *Department of Molecular Cell Biology, Institute of Medical Biology, University of Tromsø, N-9037 Tromsø, Norway*

We have discovered two introns within the nuclear rDNA of the protists *Didymium* and *Naegleria* that each encodes two group I ribozymes. These introns, DiSSU1 from *Didymium* and NaSSU1 from *Naegleria*, are the only known group I introns containing two ribozymes with different functions in RNA splicing and processing. At the RNA level, twin-ribozym-introns are organized into three distinct domains; an endonuclease coding ORF, and a small group I ribozyme (GIR1) inserted into a terminal loop (P2 in DiSSU1 and P6 in NaSSU1) of a second larger group I ribozyme (GIR2).

The GIR2 ribozymes are responsible for all the reactions involved in intron splicing, while the GIR1 ribozymes make hydrolytic cleavages at two internal processing sites (IPS1 and IPS2) located close to the 5' end of the ORFs. We hypothesize that these latter cleavages are essential to make functional mRNAs for the intron endonucleases.

The GIR1 ribozymes represent a small and unusual sub-class of group I ribozymes. GIR1 from *Didymium* (DiGIR1) is the smallest known group I ribozyme, consisting only of about 150 nucleotides. A three-dimensional computer graphic model of DiGIR1 has been constructed based on sequence alignments, site-directed mutagenesis, and chemical and enzymatical structure probing. The GIR2 ribozymes from both *Didymium* and *Naegleria* are ordinary group I ribozymes belonging to sub-group ID and IC1, respectively. Because of their unusual low salt requirements *in vitro* (*in vivo*-like conditions), intracellular stability and sequence specificity, these splicing ribozymes are promising candidates as therapeutic agents and tools, both in gene-inactivation and gene-repair. A strategy for gene therapy based on these ribozymes are under development and will be discussed.

[1] Johansen, S. and Vogt, V. M. (1994). *Cell* **76**, 725–734.
[2] Decature, W. A., Einvik, C., Johansen, S. and Vogt, V. M. (1995). *EMBO J.* **14**, 4558–4568.
[3] Einvik, C., Decature, W. A., Embley, T. M., Vogt, V. M. and Johansen, S. (1997). Submitted to *RNA*.
[4] Einvik, C., Nielsen, H., Westhof, E., Michel, F. and Johansen, S. (1997). In prep.

Does the Nuclear-encoded Mrs2 Protein Interact Directly with Group II Intron Ribozymes in Yeast Mitochondria?

Udo Schmidt[1], Irmgard Maue[1], Karola Lehmann[1], Ulf Stahl[1] and Philip S. Perlman[2]. [1]*University of Technology Berlin, Department of Microbiology and Genetics, TIB 4/4-1, Gustav-Meyer-Allee 25, 13355 Berlin, Germany;* [2]*University of Texas Southwestern Medical Center, Department of Molecular Biology and Oncology, 600 Harry Hines Blvd., Dallas, TX 75325, USA*

Although group II introns splice autocatalytically *in vitro*, genetic analysis of mitochondrial members of this group has shown that proteins are required for splicing *in vivo*. Some of the proteins are encoded by nuclear genes and presumed to participate in correct folding of the RNA.

We have focussed our interest on the *in vivo* splicing reaction of the mitochondrial group II intron aI5γ localized in the *COXI* gene of *Saccharomyces cerevisiae*. Different point mutations were introduced into a conserved substructure of the intron (domain 5) that is essential for catalysis of the ribozyme. Mitochondrially transformed yeast strains carrying one of these mutations were found to be respiratory deficient with no detectable splicing of the intron. [1,2] Splicing defects of three of the mutants were found to be suppressed by different dominant point mutations in the *MRS2* gene. The wild-type *MRS2* gene was previously shown to be essential for splicing of all four yeast group II introns and also participates in the biogenesis of mitochondria. [3]

The three suppressor *MRS2* alleles differ from the corresponding wild-type gene by three independent missense mutations in the same region of the 470 aa ORF (aa pos. 213, 230 and 232). Although no stretch of this region matches the consensus of a known RNA-binding motif the position of the mutations next to each other and their suppressor characteristics indicate that the effect of the Mrs2p on splicing is more likely to be direct than indirect. Moreover, in the C-terminal part of the protein (aa pos. 398–414) a strong hydrophilic arginine-rich motif is present that might mediate RNA binding. Further studies to identify the specific region(s) of the protein important for splicing and to demonstrate a direct RNA/protein interaction are in progress.

[1]   Boulanger *et al*. (1995). *Mol. Cell. Biol.* **15**, 4479–4488.
[2]   Schmidt *et al*. (1996). *RNA* **2**, 1161–1172.
[3]   Wiesenberger *et al*. (1992). *J. Biol. Chem.* **267**, 6963–6969.

Characterization of the Hairpin Ribozyme Using Analogues Containing 4-thiouridine (^{4S}U) and 6-thioguanosine (^{6S}G).

Tim A. Moss, James B. Murray, Chris J. Adams, John R. P. Arnold and Peter G. Stockley. *Department of Biology, University of Leeds, Leeds LS2 9JT, UK*

The hairpin ribozyme is a motif of 64 nucleotides which was identified in the minus strand of a satellite RNA of tobacco ringspot virus. Similar motifs have also been found in other satellite RNAs e.g. chicory yellow nepovirus. [1]

Dossantos *et al*. [2] show that the use of oligodeoxyribonucleotide substrate strand analogues containing a single d^{4S}U form multiple cross-links. These cross-links occur both intra- and inter-molecularly. The authors suggest this is evidence for multiple conformations of the motif in solution. Photo-crosslinking using all-RNA substrates containing a single substitution by a photoactivatable group (^{4S}U [3] or ^{6S}G [4]) has been explored. These substrate analogues show either a single cross-link, or the complete absence of any cross-linked species. Cross-link yields in excess of 50% are reproducibly observed with ^{4S}U substituted strands and 10% using ^{6S}G. A single cross-linked species from any single substitution suggests that the all-RNA hairpin ribozyme has a single conformation in solution. The contrast between these results and those produced using DNA substrate analogues could be due to the distinct base-pairing conformations adopted by RNA and mixed DNA–RNA double helices. Kinetics of the substrate analogues show that the substitutions affect the rate of cleavage by more than ten-fold. It is therefore possible that the conformation adopted by the analogue–enzyme complex may not accurately reflect that seen in the wild-type system.

[1]   Symons, R. H. (1994). *Current Opinion in Structural Biology* **4**, 322–330.
[2]   Dossantos, D. V., Vianna, A. L., Fourrey, J. L. and Favre, A. (1993). *Nucleic Acids Res.* **21**, 201–207.

[3]   Adams, C. J., Murray, J. B., Arnold, J. R. P. and Stockley, P. G. (1994). *Tetrahedron Letters* **35**, 765–768.
[4]   Adams, C. J., Murray, J. B., Farrow, M. A., Arnold, J. R. P. and Stockley, P. G. (1995). *Tetrahedron Letters* **36**, 5421–5424.

A Ribozyme that Catalyses 2′-5′-phosphodiester-bond Formation. A. Jenne and M. Famulok. *Institut für Biochemie, Ludwig-Maximilians-Universität München, Feodor-Lynen-Str. 25, 81377 München, Germany*

In vitro selection techniques by which vast libraries of nucleic acids are screened for certain functionalities have proven to be a powerful tool for isolating ribozymes with properties that might have implications for a hypothetical prebiotic 'RNA world'. [1] If modern translation has evolved in this 'RNA world', RNA must have had the ability to recognize amino acids and to catalyse several chemical reactions: amino-acid activation, RNA aminoacylation, aminoacyl-transfer and peptide synthesis. [2]

In this study, we aimed toward the selection of ribozymes with aminoacyl-transfer activity. A pool of 10^{15} different RNAs was incubated with the 2′/3′-OAMP-ester of N-biotinylated phenylalanine (BPA). After 13 cycles of selection, three classes of active sequences were isolated, all catalysing a reaction in which the ribozyme is linked covalently to the biotin anchor. Kinetic analysis of one of the ribozymes (clone 11) revealed that it obeys standard Michaelis-Menten kinetics for a single turnover reaction with a single saturable binding site ($k_{cat} = 0.22$ min^{-1}, $k_M = 145\,\mu$M). Characterization of the chemical reaction catalysed, however, revealed that instead of the aminoacyl-transfer reaction, the entire molecule BPA is attached to the ribozyme. After cleaving the labile 2′/3′-OAMP-ester with a sulphur nucleophile, MALDI-TOF analysis revealed that AMP was still linked to the ribozyme. RNAse sequencing of the modified ribozyme and mutation analysis strongly suggest that AMP is bound via its 5′-phosphate to a certain 2′-hydroxy group within the RNA. The newly formed 2′-5′-phosphodiester bridge resembles a branch point as found, for example, during group II splicing.

Of particular interest is the mechanism by which the clone 11 ribozyme mediates the phosphodiester-bond formation. The chemistry of the reaction catalysed strongly indicates that the ribozyme is capable of performing the condensation of a phosphate group with an alcohol, without the requirement of activated monomers. This might be achieved by a catalytic-triad-like mechanism in which the attacking 2′-hydroxyl is deprotonated, whereas a phosphorus OH-group of BPA becomes protonated to enable the leaving of a water molecule. A reaction in which RNA utilizes nucleoside-monophosphate as substrates for phosphordiester-bond formation might be relevant for ribozymes as possible predecessors of modern time ligases.

[1]   Joyce, G. F. (1996). *Curr. Biol.* **8**, 965–967.
[2]   Hager, A. J., Pollard, J. D. and Szostak, J. W. (1996). *Chem. Biol.* **3**, 717–725.

Inhibition by Uridine but not Thymidine of p53-dependent Intestinal Apoptosis Initiated by 5-fluorouracil: Evidence for the Involvement of RNA Perturbation Initiating a p53 Response.

John A. Hickman[4], D. Mark Pritchard[1,2,4], Alastair J. M. Watson[2], Christopher S. Potten[1] and Ann L. Jackman[3]. [1]*Cancer Research Campaign Department of Epithelial Cell Biology, The Paterson Institute for Cancer Research, Christie Hospital Trust, Wilmslow Road, Manchester M20 9BX;* [2]*Department of Medicine, University of Manchester, Hope Hospital, Eccles Old Road, Salford M6 8HD;* [3]*Institute of Cancer Research, Belmont, Sutton, Surrey, SM2 5NG;* [4]*Molecular and Cellular Pharmacology Group, School of Biological Sciences, G38 Stopford Building, University of Manchester, Manchester M13 9PT, UK*

The epithelia from the crypts of the intestine are exquisitely sensitive to metabolic perturbation and undergo cell death with the classical morphology of apoptosis. Administration of $40\,\mathrm{mg\,kg}^{-1}$ 5-fluorouracil to BDF-1 p53 +/+ mice resulted in an increase in p53 protein at cell positions in the crypts which were also those subjected to an apoptotic cell death. In p53 −/− mice apoptosis was almost completely absent, even after 24 h. 5-fluorouracil is a pyrimidine antimetabolite cytotoxin with multiple mechanisms of action, including inhibition of thymidylate synthase, which gives rise to DNA damage, and incorporation into RNA. The inhibition of thymidylate synthase can be increased by co-administration of folinic acid and can be abrogated by administration of thymidine. The incorporation of 5-fluorouracil into RNA is inhibited by administration of uridine. p53-dependent cell death induced by 5-fluorouracil was only inhibited by administration of uridine. Uridine had no effect on the apoptosis initiated by 1 Gy of γ-radiation. Although thymidine abrogated apoptosis induced by the pure thymidylate synthase inhibitor Tomudex, it had no effect on 5-fluorouracil-induced apoptosis and coadministration of folinic acid did not increase apoptosis. The data show that 5-fluorouracil-induced cell death of intestinal epithelial cells is p53 dependent and suggests that changes in RNA metabolism initiate events culminating in the expression of p53.

The N-terminus of Eukaryotic Translation Elongation Factor-3 (EF-3) Binds to 18 S rRNA.

Richard R. Gontarek and Cathy Prescott. *SmithKline Beecham Pharmaceuticals, RNA Research Group, Collegeville, PA 19426-0989, USA*

Elongation factor-3 (EF-3) is a fungal-specific translation factor that is absolutely required for protein synthesis. The protein consists of a single polypeptide chain of 125 kDa, and exhibits strong ribosome-dependent ATPase and GTPase activities. While its exact role remains unclear, EF-3 reportedly participates in aa-tRNA binding to the ribosomal A-site and in release of deacylated tRNA from the E-site. EF-3 exhibits sequence homologies that may predict structural domains critical for its role in protein synthesis. [1] These include several Walker motifs found in ATP-binding proteins, a C-terminal region homologous to aa-tRNA synthetases, and a highly charged C-terminal tail. In addition, an extended domain at the N-terminus of EF-3 is homologous to *E. coli* ribosomal protein S5. It has been reported that S5 binds to 16 S rRNA; therefore the aim of this work is to explore whether EF-3 binds to 18 S ribosomal RNA from *S. cerevisiae*.

A portion of the N-terminus of *S. cerevisiae* EF-3 (spanning the S5 homology region, amino acids 98–388) has been cloned with a 6xHis-tag and expressed in bacteria. This 35 kDa protein was purified under denaturing conditions via Ni-NTA agarose affinity chromatography. UV cross-linking experiments revealed that the N-terminal EF-3 protein (N-term EF-3) forms a covalent cross-link to 18 S rRNA. Cross-linking of N-term EF-3 to radiolabelled 18 S rRNA is competed by unlabelled 18 S rRNA but not by non-specific competitor RNAs. Therefore, the interaction between N-term EF-3 and 18 S rRNA is apparently specific. A filter-binding assay to further examine this interaction confirmed the results of the UV cross-linking, and also established that the interaction was relatively rapid and has a $Kd = 0.2 \, \mu M$. The nucleotide and amino acid residues involved in this interaction, as well as the functional significance of the N-term EF-3/18 S rRNA interaction, remain to be determined.

[1] Belfield, G. P. and Tuite, M. F. (1993). *Mol. Microbiol.* **9**, 411–418.

Whole Cell Selection of an RNA Fragment that Binds Spectinomycin. Therese Sterner, George Thom, Hu Li and Cathy Prescott. *SmithKline Beecham Pharmaceuticals, RNA Research Group, Collegeville, PA19426-0989, USA*

The translation apparatus is essential to all living cells and represents one of the major targets for antibiotics. An understanding of the mechanism of drug action requires detailed knowledge of interactions at the molecular level. Such information is clearly difficult to obtain when faced with the size and complexity of the ribosome. Several lines of evidence support the notion that the ribosome can be fragmented into smaller, functional subdomains that retain organizational and ligand-binding properties characteristic of the intact particle. [1] For example, the *E. coli* 16 S ribosomal RNA is organized into three major domains and rRNA fragments representing each of these subdomains can reassemble with specific subsets of ribosomal proteins. [2,3] A 16 S rRNA fragment encompassing the 3′ domain (nucleotides 923–1542) can be reconstituted together with eight ribosomal proteins, forming a compact particle that resembles the head of the 30 S subunit. This particle retains the property of binding the antibiotic spectinomycin that specifically protects the N-7 position of G1064 from attack by dimethyl sulphate in both 30 S subunits and the sub-particle. [3,4] Accordingly, we reasoned that RNA molecules expressed *in vivo*, may fold in such a manner as to mimic a drug-binding site present on the intact ribosome. If expressed at sufficient levels, the RNA would sequester the antibiotic thereby permitting the continued function of the ribosome and consequently allow the cell to survive in the presence of the drug. Evidence to support the 'RNA fragment rescue' concept is provided, following the selection and characterization of RNA fragments that confer resistance to the antibiotic spectinomycin.

[1] Schroeder, R. (1994). *Nature* **370**, 597–598.
[2] Noller, H. F. and Woese, C. R. (1981). *Science* **212**, 402–411.
[3] Samaha, R. R., *et al*. (1994). *Proc. Natl. Acad. Sci.* **91**, 7884–7888.
[4] Moazed, D. and Noller, H. F. (1987). *Nature* **327**, 389–394.

In Vitro Selection of a Viomycin Binding RNA Pseudoknot.

Mary G. Wallis[1], Barbara Streicher[1], Herbert Wank[1], Uwe von Ahser[1], Michael Famulok[2] and Renee Schroeder[1]. [1]*Institute of Microbiology and Genetics, University of Vienna, Dr. Bohrgasse 9, Vienna, A-1030 Austria;* [2]*Institute of Biochemistry, Genzentrum of the Ludwig-Maximilian-University of Munich, Wurmtalstrasse 221, Munich, D-81375 Germany*

Employing *in vitro* selection we have isolated viomycin-binding RNA molecules from a pool of 1015 different random-sequence RNAs. Sequence comparison of individual RNAs revealed that one continuous highly conserved region (14nt) was shared by more than 90% of the selected molecules. Structural probing together with footprinting experiments (chemical modification and Pb2+ cleavage) demonstrated that this conserved sequence region forms a stem-loop structure and harbours the antibiotic binding site. Alkaline hydrolysis interference studies, used to determine the minimal sequence requirements, revealed that the stem-loop motif is not sufficient for binding. Mutational analysis identified 3′ terminal sequences base pairing with part of the consensus motif resulting in the formation of a pseudoknot. Interestingly, three natural RNA target sites (in 16 S rRNA of *E. coli*, in group I introns and the human hepatitis delta virus ribozyme), which are inhibited by viomycin, form pseudoknots at the antibiotic binding site. This observation suggests that viomycin requires a complex RNA structure involving tertiary structural elements for recognition and explains why viomycin enforces RNA–RNA interactions.

In Vitro Selection of Dopamine RNA Ligands.

Cecilia Mannironi[1], Alessia Di Nardo[1], Paolo Fruscoloni[1]* and G. P. Tocchini-Valentini[1,2]. [1]*Institute of Cell Biology, CNR, Viale Marx 43, 00137 Rome, Italy;* [2]*Department of Biochemistry and Molecular Biology, University of Chicago, Chicago IL 60637, USA.* *Corresponding author*

RNA aptamers that specifically bind dopamine have been isolated by *in vitro* selection from a pool of 3.4×10^{14} different RNA molecules. One aptamer (dopa2), which dominated the selected pool, has been characterized and binds to the dopamine affinity column with a dissociation constant of $2.8 \mu M$. The specificity of binding has been determined by studying binding activity of a number of dopamine-related molecules, showing that the interaction with the RNA might be mediated by the hydroxyl group at position 3 and the proximal carboxylic chain in the dopamine molecule. The binding domain was initially localized by boundary experiments. Further definition of the dopamine binding site was obtained by secondary selection on a pool of sequences derived from a partial randomization of the dopa2 molecule. Sequence comparison of a large panel of selected variants revealed a structural consensus motif among the active aptamers. The dopamine binding pocket is built up by a long-range interaction between two stem and loop motifs, creating a stable framework in which five invariant nucleotides are precisely arrayed. Minimal active sequence and key nucleotides have been confirmed by the design of small functional aptamers and mutational analysis. Enzymatic probing shows that the RNA undergoes a conformational change upon ligand binding that stabilizes the proposed tertiary structure.

Helix 1 of *E. coli* 23 S rRNA is Essential for Ribosomal 50 S Subunit Assembly. Aivar Liiv and Jaanus Remme. *Department of Molecular Biology, Institute of Molecular and Cell Biology, Tartu University, Riia 23, EE2400 Tartu, Estonia*

In the ribosomal RNA precursors the spacer sequences bracketing mature 16 S and 23 S rRNA are base-paired to form long helices (processing stems). In the mature 23 S rRNA the processing stem is continued and forms helix 1. We have constructed a series of deletion and substitution mutations in the processing stem and in the helix1 on the plasmid copy of 23 S rRNA gene. Mutant rRNA genes were expressed in rnc⁺ and rnc⁻ (RNase III) strains. Incorporation of the plasmid-encoded 23 S rRNA into 50 S subunits, 70 S ribosomes, and polysomes was determined by RNA sequencing. Functional activity of the plasmid borne ribosomes was analysed during *in vitro* translation. Results:

(i)   Deletions and substitutions in both strands of the helix 1 lead to the loss of plasmid derived 50 S formation. Double mutations restoring helix 1 were assembled into functional 50 S subunits.

(ii) Deletions in 5′ and 3′ halves of the processing stem reduced the ability of 23 S rRNA to form 50 S subunits. Increasing deletions in the 3′ half were accompanied by the gradual loss of 50 S assembly. However, even the complete removal of the 3′ strand of the processing stem did not completely abolish the 50 S assembly.

(iii) Most of the mutations exhibited similar effect on the 50 S assembly in both rnc⁺ and rnc⁻ strains except mutations in the helix 1. Alteration of the helix 1 was partially compensated by the presence of the processing stem in the assembled ribosomes in the rnc⁻ strain indicating that the helix 1 is not essential for the functioning of the assembled 50 S ribosomal subunits.

We conclude that the helix 1 is the main ribosome assembly determinant of 23 S rRNA. Processing stem facilitates assembly, however it is not essential. Base-pairing between both ends of 23 S rRNA is a key event during 50 S assembly.

An Elongated Hairpin Structure Controls the Ribosome Shunt on the CaMV 35 S RNA Leader *in Vivo* and *in Vitro*; Alternative Coexisting Conformations Might Fulfil Specialized Functions. Maja Hemmings-Mieszczak and Thomas Hohn. *Friedrich Miescher-Institut, Maulbeerstrasse 66, 4058 Basel, Switzerland*

The cauliflower mosaic virus (CaMV) 35 S RNA functions as both messenger and pre-genomic RNA under the control of its 600 nts leader, which contains regulatory elements involved in splicing, polyadenylation, translation, reverse transcription, and probably also packaging. The structure of the leader has been characterized theoretically and experimentally. [1] The predicted conformation, a low-energy elongated hairpin, base-pairing the two halves of the leader, with a cross-like structure at the top, is strongly supported by enzymatic probing, chemical modification and phylogenetic comparison. The elongated hairpin is stabilized by strong base-pairing between the ends of the leader, including regions which are important in allowing translation downstream of the leader *via* the ribosome shunt mechanism. [2] Mutations destabilizing the base of the elongated hairpin decrease the ribosome shunt and the expression of the CAT-reporter gene in wheat germ *in vitro* translation systems and in transiently transfected plant protoplasts. Introduction of compensatory mutations into the second half of the leader restores both the elongated hairpin structure and translation efficiency.

The 35 S RNA multifunctional leader can adopt additional alternative conformations. We find that each of the possible conformations reveals a distinct melting curve that can be influenced by mutations. At low ionic strength, temperature gradient gel electrophoresis detects a single elongated hairpin structure that melts at 32–38°C. The main transition step is irreversible and leads to a drastic retardation of gel mobility due to formation of a big internal loop closed by stable base-pairing between the ends of the leader. Two higher order conformations of low electrophoretic mobility, a long-range pseudoknot connecting central and terminal parts of the leader, and a dimer, which were detected at high ionic strength and at low temperature, differ in stability. An irreversible transition of the pseudoknot occurs at 25°C leading to the elongated hairpin conformation. The structural transition of the dimer occurs at the same temperature as the main retardation transition of the monomer and indicates that dimerization probably involves sequences that otherwise stabilize the elongated hairpin structure of the monomer. The elongated hairpin, long-range pseudoknot and dimer might coexist and fulfil specialized functions.

[1] Hemmings-Mieszczak, M., Steger, G. and Hohn, T. (1997). *Journal Mol. Biol.* (in press).

[2] Fütterer, J., Kiss-László, Z. and Hohn, T. (1993). *Cell* **73**, 789–802.

Three-dimensional Structure of the 70 S Ribosome in Different Functional States.

Holger Stark[1], Elena V. Orlova[1], Jutta Rinke-Appel[2], Marina Rodnina[3], Wolfgang Wintermeyer[3], Richard Brimacombe[2] and Marin van Heel[1]. [1]*Imperial College, Department of Biochemistry, London SW7 2AY, UK;* [2]*Max-Planck-Institut für Molekulare Genetik, D-14195 Berlin, Germany;* [3]*Institut für Molekular-biologie, Universität Witten/Herdecke, D-58453 Witten, Germany*

Electron cryomicroscopy is playing an increasingly important role in structural studies of large biological macromolecules. The angular reconstitution technique [1] allows one to exploit the random orientations that the macromolecules can assume in a homogeneous embedding matrix in order to calculate the three-dimensional structure of the macromolecule. A large number of structures have thus been elucidated in the last several years to a resolution of 15–30 Å, including: the skeletal muscle Ca^{2+}-release channel in its closed [2] and open [3] states; the giant haemoglobin of the common earthworm *Lumbricus terrestris* [4]; the haemocyanin of the keyhole limpet snail 'KLH1'; [5] and the *E. coli* 70 S ribosome. [6]

The particles in their native, hydrated state are frozen rapidly by being plunged into liquid ethane, thus generating samples of randomly oriented particles in a homogeneous matrix of vitreous ice. By 'freezing' the macromolecules into specific functional states, one can study a complete series of structural transitions in a molecular assembly. We have recently studied the structure of the 70 S ribosome in both its pre- and its post-translocational states. [7] In the pre-translocational state tRNA molecules are discernible in the A and P sites of the ribosome; whereas in the post-translocational state tRNA molecules can be seen in the P and E sites. We now have also frozen the ribosome in a different functional state of the elongation cycle, in which the ternary complex (consisting of EF-Tu, GTP and amino-acyl tRNA) is attached to its ribosomal binding site (manuscript submitted for publication). The structure of the ternary complex, within the electron microscopical reconstruction, is in excellent agreement with its known X-ray structure. The main difference between the electron microscopic density and the X-ray structure is in the angle between the EF-Tu and the tRNA, which difference probably reflects a conformational change due to codon recognition.

[1] van Heel, M. (1987). *Ultramicroscopy* **21**, 111–124.
[2] Serysheva, I. I. *et al.* (1995). *Nature Struct. Biol.* **2**, 18–24.
[3] Orlova, E. V. *et al.* (1996). *Nature Struct. Biol.* **3**, 547–552.
[4] Schatz, M. *et al.* (1995). *J. Struct. Biol.* **114**, 28–44.
[5] Orlova, E. V. *et al.* (1997). submitted.
[6] Stark *et al.* (1995). *Structure* **3**, 815–821.
[7] Stark *et al.* (1997). *Cell* **88**, 19–28.

Characterization of Metal (Eu^{3+}) Binding Sites in Ribosomes.

Silke Dorner and Andrea Barta. *Institute of Biochemistry, University of Vienna, Vienna Biocenter, Dr. Bohrgasse 9, A-1030 Vienna, Austria*

The integrity as well as all functional aspects of ribosomes are crucially dependent on magnesium ions, but virtually nothing is known about specific metal binding sites in this important enzymatic complex. We therefore want to study metal binding pockets in rRNA by metal-induced specific hydrolysis of RNA in ribosomes. Metal hydroxides bound to highly structured RNA are able to cleave a structurally adjacent phosphodiester bond. Experiments using lead for site-specific cleavage of ribosomal RNA have already indicated at least three strong metal binding sites in the ribosome. [1]

Now we have done similar experiments with lanthanide metal ions because they were often used to probe alkaline earth metal binding sites in proteins and RNA due to their similarity to these ions and their easily examined spectral properties. In most cases lanthanide ions could substitute for calcium or magnesium ions in macromolecules without destroying their biological function. X-ray crystallographic studies in yeast tRNA[Phe] have shown that lanthanide ions are able to displace magnesium ions from strong binding sites. [2]

70 S ribosomes were incubated with $100 \mu M$ $EuCl_3$ for various times at pH 7.4 and 37°C. Ribosomal RNA was isolated and used as template for primer extension analysis to locate the nucleotides where cleavage had occurred. We found several cleavages on 16 S rRNA and on 23 S rRNA. The locations of these sites in two-dimensional models of these RNAs are shown. Some of the cleavages occur in regions which were implicated in tRNA binding and antibiotic interactions.

Interestingly, we observed an influence of erythromycin and chloramphenicol (both of them are known to inhibit peptide bond formation) on cleavage sites situated in the peptidyltransferase region (central loop of the domain V) of the 23 S RNA. The effects of binding of tRNA and antibiotics on the various europium cleavage sites in the 16 S and 23 S ribosomal RNA will be discussed.

[1]   Winter *et al.* (1997). *Nucleic Acids Res.* Accepted.
[2]   Jack *et al.* (1977). *J. Mol. Biol.* **111**, 315–328.

A Ribonuclease Specific for Inosine-containing RNA: a Potential Role in Antiviral Defence? A. D. J. Scadden and Christopher W. J. Smith. *Department of Biochemistry, University of Cambridge, Cambridge CB2 1QW, UK*

RNA transcripts in which all guanosine residues are replaced by inosine are degraded at a highly accelerated rate when incubated in extracts from HeLa cells, sheep uterus or pig brain. We have partially purified and characterized a novel ribonuclease, referred to as I-RNase, that is responsible for the degradation of inosine-containing RNA. I-RNase is Mg^{2+} dependent and specifically degrades single-stranded inosine-containing RNA. Comparison of the K_m of the enzyme for inosine-containing RNA with the K_i for inhibition by normal RNA suggests a $\sim$300 fold preferential binding to I-RNA, which can account for the specificity of degradation. The site of cleavage by I-RNase is non-specific; I-RNase acts as a $3' \rightarrow 5'$ exonuclease generating $5'$-NMPs as products. The presence of alternative unconventional nucleotides such as 5-Bromo-uridine in RNA does not result in degradation unless inosine residues are also present. We show that I-RNase is able to degrade RNAs that have previously been modified by the RED-1 dsRNA adenosine deaminase (dsRAD). dsRADs destabilize dsRNA by converting adenosine to inosine, and some of these enzymes are interferon inducible. We therefore speculate that I-RNase in concert with dsRAD may form part of a novel cellular anti-viral defence mechanism that acts to degrade dsRNA.

Compact Structures of CUG Repeats in RNA, and Abortive Transcription Mediated by Trinucleotide Repeats: Potential Implications for Human Genetic Diseases.

Philip Pinheiro[1], Garry Scarlett[1], Anna Murray[2], Tom Brown[3], Sarah Newbury[1] and James McClellan[1,*]. [1]*Biophysics Laboratories, School of Biological Sciences, University of Portsmouth, St Michael's Building, White Swan Road, Portsmouth PO1 2DT, UK;* [2]*Wessex Regional Genetics Laboratory, Salisbury Health Care NHS Trust, Salisbury District Hospital, Salisbury, Wiltshire SP2 8BJ, UK;* [3]*Lab 5005, Medical and Biological Sciences Building, University of Southampton, Boldrewood, Bassett Crescent East, Southampton SO16 7PX, UK. *Corresponding author*

Triplet repeats that cause human genetic diseases have been shown to exhibit unusual compact structures in DNA, and in this paper we show that similar structures exist in CNG RNA. CUG and control RNAs were made chemically and by *in vitro* transcription. We find that CUG RNAs migrate anomalously fast on native gels, compared with control oligos of similar base composition. Additionally we demonstrate that T7 RNA polymerase exhibits frequent termination when transcribing templates that encode triplet repeat RNA. We discuss the possible contribution of CNG RNA structure to normal and abnormal function of human genes.

Quantification of Gene Expression Using Scintillating Microplates.

M. K. Kenrick[1], D. W. Harris[2], R. J. Pither[1], D. A. Jones[2] and J. G. Anson[1*]. [1]*Amersham International plc, Forest Farm, Whitchurch, Cardiff CF4 7YT, UK;* [2]*Pharmacia & Upjohn Inc, 301 Henrietta Street, Kalamazoo, MI 49001, USA. *Corresponding author*

A plate-based *in situ* hybridization assay has been developed, utilizing Amersham™ Cytostar-T™ scintillating microplates, which has the sensitivity to reliably detect specific mRNA transcripts at the level of 20–30 copies per cell. Anti-sense riboprobes specific for *c-fos* (*c-fos₂₃₆*) and glyceraldehyde-3-phosphate dehydrogenase (GAPDH), as well as a non-homologous vector-derived control probe, were labelled to high specific-activity ($> 4 \times 10^9$ cpm µg^{-1}) with [α-^{33}P]UTP. These probes were used to compare mRNA levels in rat A10 cells after stimulation with fetal calf serum or PDGF. Maximal *c-fos* induction was shown to occur following stimulation of quiesced cells with 30 ng ml^{-1}

PDGF or 10% FCS, corresponding to a signal from the c-fos_{236} probe of 350 cpm. The non-homologous control background of 50 cpm and the GAPDH signals of 2250 cpm were independent of stimulation with PDGF or serum. Using PDGF, at 30 $ng\,ml^{-1}$, quiesced cells were stimulated at various times to provide an induction time-course for c-fos mRNA. Expression of the c-fos transcript peaked at 30 min and decreased to less than 50% of this by 90 min; a return to background level expression was apparent after 3 hours. The relative simplicity and improved sensitivity of the Cytostar-T microplate assay in comparison to Northerns, clearly makes it amenable to automation, and thus the potential exists to configure high-throughput screening assays for the detection of mRNA.

Analogues of Hoechst 33258 Bind to Transfer RNA with 1:1 Stoichiometry. S. E. Sadat Ebrahimi, Amanda N. Wilton and Kenneth T. Douglas. *School of Pharmacy and Pharmaceutical Sciences, University of Manchester, Manchester, M13 9PL, UK*

Early studies indicated netropsin and distamycin to bind strongly to the DNA minor groove but weakly to RNA (probably electrostatically to the phosphate backbone). [1] However, classical DNA minor-groove ligands [2] can interact strongly with RNA duplexes based on T_m studies. However, these studies gave no indication of RNA-specifying interactions and the grooves of DNA and RNA differ in many respects. An X-ray diffraction study of tRNA with netropsin and distamycin showed specific hydrogen bonds and electrostatic interactions with phosphates [3] but binding strengths were not measured. Hoechst 33258 (**1**) binds strongly to B-DNA in AT-rich regions but the absolute sequence specificity is low, with multiple sites on calf thymus DNA. We report here specific binding to a unique site on tRNA of **2** and **3**.

1:   X = Z = H : Y = OH
2.   X = Y = OCH_3 : Z = H
3.   X = Y = Z = OH

In the absorption spectra of mixtures of tRNA (brewers' yeast) and 3 the λ_{max} value of 3 shifted from 326 nm with increased tRNA, becoming $\sim$338 nm for a 1:1 molar mixture. Spectra were recorded for a series of tRNA:3 mixtures, varying the molar ratio of the components but keeping a constant total concentration of the components. The absorbance at 360 nm for each ratio, corrected for unbound 3 or tRNA assuming 1:1 tight binding, plotted against the mole fraction of 3 as a Job plot, showed 1:1 complex formation and tight binding. Addition of tRNA to a solution of 2 led to fluorescence quenching. A sample of 2 in 0.1 M sodium phosphate buffer, pH 7.80 was excited at 338 nm and its fluorescence emission spectrum recorded between 300 and 600 nm after each addition of a series of aliquots of $tRNA^{Phe}$ (20 μl from 100 μM stock). A plot of the change in the fluoresence intensity, corrected for dilution, *versus* $[tRNA^{Phe}]$ added, analysed for single-site binding (Fluorescence(F) $= F_{max}[tRNA^{Phe}]/(K_{diss} + [tRNA^{Phe}]))$, gave $K_{diss} = 14.8 + 0.9$ μM, with evidence of other, much weaker binding site(s).

The environment of these ligands on the RNA clearly differs in detail from that on DNA as the fluorescence of Hoechst 33258 is highly enhanced on binding to the DNA minor groove [4] but quenched by tRNA. The molecular frameworks of 2 and 3 and their specific sequestration of tRNA (selectively occupying a unique run of 4–5 base pairs out of approximately 75 bases) provides a new lead to understand some aspects of ligand:RNA recognition at an atomic level.

[1] Zimmer, C., Reinert, K. E., Luck, G., Wahnert, U., Lober, G. and Thrum, H. (1971). *J. Mol. Biol.* **58**, 329–348.
[2] Pilch, D. S., Kirolos, M. A., Liu, X., Plum, E. and Breslauer, K.J. (1995). *Biochem.* **34**, 9962–9976.
[3] Rubin, J. and Sundaralingam, M. (1984). *J. Biomol. Struct. Dynamics* **2**, 165–174.
[4] Parkinson, J. A., Barber, J., Douglas, K. T., Rosamund, J. and Sharples, D. (1990). *Biochem.* **29**, 10181–10190.

Index

Numbers in *italics* refer to illustrations